Potentiality of Indian Reservoirs of Tropical and Subtropical Regions in the Context of Aquaculture

Potentiality of Indian Reservoirs of Tropical and Subtropical Regions in the Context of Aquaculture

Author
P. Ray

2009
DAYA PUBLISHING HOUSE
Delhi - 110 035

ISBN10 81-7035-561-3
ISBN13 978-81-7035-561-8

Published by : **Daya Publishing House**
1123/74, Deva Ram Park
Tri Nagar, Delhi - 110 035
Phone: (011) 27383999
Fax: (011) 23260116
e-mail : dayabooks@vsnl.com
website : www.dayabooks.com

Showroom : 4760-61/23, Ansari Road, Darya Ganj,
New Delhi - 110 002
Phone: 23245578, 23244987

Laser Typesetting : **Classic Computer Services**
Delhi - 110 035

Printed at : **Chawla Offset Printers**
Delhi - 110 052

PRINTED IN INDIA

Acknowledgements

In my earlier publication on *"Ecological Imbalance of the Ganga River System: Its Impact on Aquaculture"* envisaged me on several occasions on the possibility and extent of aquaculture in Reservoirs/lakes linked with rivers, which are normally meant for storing water for wide use in agriculture, power generation, water supply to various industries and cities etc. I discussed the matter with my friend and colleague Dr. P. R. Sen, Retired principal investigator, CIFA, (ICAR) who welcomed my ideas by encouraging me and by offering some valuable suggestions time to time, such write up will not only help to limnologists but also to various concerned states for developmental work in their Reservoirs/lakes located in tropical and sub-tropical regions.

My special thanks are also due to Doctors V. G. Jhingran (1975), A.G. Jhingran (1991), all retired Directors [CIFRI, Barrackpore, (West Bengal)], V.V. Sugunan, Assistant Director General (IFY) (ICAR), Pusa, New Delhi and to Dr. G.P. Dubey [Director, M.P. and Member, National Commission on Agriculture Part-VIII (Fisheries) Government of India, Ministry of Agriculture and Education, New Delhi, 1976] for using some relevant data from their books while preparing this document.

Finally, I owe greatest to my dear nephews Dr. Sujit Roy for his constructive suggestions and to Bidyut Roy for his constant inspiration until the completion of the work.

P. Ray

Preface

With the growing population in the country the demand for water has considerably increased for running the industries, agriculture, power generation, domestic purposes, flood control and other water oriented activities for human comfort. River water alone cannot meet such huge demand of water round the year and this has resulted in and accelerated growth of reservoirs or dam construction (mostly after independence) to hold enough of water during the wet months from where considerable volume of water is drawn round the year for use in various purposes indicated. This results considerable fluctuation in the water level. Thus the stability in water level is remote possible. Consequent to the draw down of water large bays of the reservoir get exposed and the process is continued till monsoon breaks again. Added to it draughts or late monsoon deteriorates the ecosystem further.

Deforestation has become essential for human habitation. It is reported that the assessment made in 1987 (based on imagery, 1981-83) the country had forest cover of about 6,42,041 sq.km. which is reduced to 6,40,134 sq.km. in 1989. This results erudition of erosion volume of silt and other organic materials which was brought into the reservoir bed year after year. As a result the bed is gradually reduced. A number of large and small reservoirs are highly irregular in their shore line which indicates presence of shallow littoral zones supposed to be productive. Random sampling of volume development index (high and low) of reservoirs shows that the value ranges from 1.5 to as high as 13.5. Contrary to this concave lake basin is also recorded (0.53-0.89, where volume development index is less than 1).

Bera, 1984 (based on 1972 irrigation report) has shown actual life span (in year) of some reservoirs based on assumed silt precipitation which shows the calculated life span as 150 years in place of assumed life span as 212 years for Maithon (14021 ha), 138 years against 216 years for Panchet (7511 ha), 200 years against assumed 872 years for Mayurakshrai (6734 ha), 291 years against assumed 403 years for Bhakra also known at Govind Sagar (16867 ha), 147 years against assumed 386 years for Hirakud (71963 ha), 350 years against assumed 930 years for Gandhi Sagar (66,000 ha) and 245 years against assumed 311 years for Tungabhadra Reservoir (37914 ha). With the passing of time such silt precipitation may further accelerate in all the reservoirs depending upon the soil character of the catchments area, when the actual purpose of the reservoir/dam construction may be vitiated.

The Geographic location affects the metabolism of a reservoir through nutrient supply, shape of basin and efficiency with which the climatic factors able to act in the dynamic exchange. In addition the hydrological cycle (inflow and out flow) also play important role to affect the final productivity.

Normally reservoirs are constructed on different river systems. For example the main reservoirs are located in the Ganga river system, Brahmaputra river system, Mahanandi river system, Krishna river system, Godavari river system and Western Coast etc.

Transformation from river to reservoir causes reduced flow in the reservoir, heavy silt precipitation, gradual change in the water quality etc. The flora and faunal composition with species (qualitative and quantitative) of the main river is gradually changed which is more suited to the reservoir. The benthos adopted in the deeper zone is different from the shallow zone or from lotic condition of the river. Benthic invertebrates are adversely affected, reduces insect diversity density and species richness. Fishes are affected by alteration, siltation; degrade spawning grounds, cause behavioural change in spawning fishes, increase egg mortality, decrease larval growth and reduces magnitude of flow down-steam for most of the dry seasons. The fisheries of reservoirs are mainly influenced by the fishes inhabiting in the rivers joining the reservoir. Broadly, the classification of species inhabiting in different reservoirs are: Indian major carps, Mahaseers, minor carp including snow trout and peninsular carps, large catfishes, feather backs, air breathing catfishes, murrels, weed fishes, exotic fishes etc.

The fishes of Indian reservoirs is constituted of both indigenous and transplanted fish populations, Among the former, North Indian Reservoirs, both as naturally occurring and stocked fishes. In addition to these they also harbour *Labeo bata, Puntius sarana, P. Chaunio* and *Cirrhinus reba*. However, in Govind Sagar, the major carps constituted very low percentage of total fisheries, as the dominant fisheries comprised *Tor-putitora Labeo dero and Schlzothorax plagiostomus*.

In peninsular reservoirs, the indigenous fishes forming the commercial fisheries are *Cirrhinus cirrhosa, C. reba, Labeo Kuntius, L. fim-briatus, Puntius dobius, P. Sarana, P. carnaticus, P. colus, P. dobsoni, P. hexugoralopis, Tor tor, Thynnichthyes, Sand khol*, and *Osteobrama Vigorsii*. In addition, Gangetic major carps are also stocked.

Among large catfishes, the Indian reservoirs contain *Wallago atto Aorichthys seenghela, Mistus aor, Silonia silondia* and *Pangasius pangasius, Silonia childreni and Pseudotropies takree* have established in Tungabhadra Reservoir (David, 1969) and *Mystus vittatus* in Poondi reservoir (Chacko, 1969). Indian reservoirs also horbour a sizeble population of trash fishes like *Ambassis nama, Esomus danrica, Aspidoparia, morar, Ambypharyngodon mola, Puntius sophore, Pautius, P. ticto. Oxygaster bacaila,*

Laubuca laubuca, Barilius-barilia, Osteobrama Cotio and *Gadusia Chapra*. Most of these trash fishes compete for food with Indian Major carps tending to reduce the overall productivity of the reservoir.

Reservoirs in North India belt come under the monomictic lakes with strong thermal stratifications during summer. But during winter both wind action and monsoon floods break the summer stratification (Rihand and Govind Sagar). In the peninsular reservoirs on the other hand generally show absence of thermal stratification (Nagarjuna Sagar and Bhavani Sagar), circulation of water brings locked chemical nutrients of the tropholytic zone upto trophogenic zone and facilitates fixation and utilization of energy. On the other hand reservoirs like Bhavanisagar, Nagarjuna Sagar and Govind Sagar have shown strong oxygen decline in the tropholytic layers and distribution of oxygen in Rhino grade in nature. These reservoirs can therefore be said high energy reserve at the bottom' and as such can come under the category of high productive reservoir. On the other hand Rihand showed a near uniform distribution of oxygen from surface to bottom (Orthograde) and can be put under low productive reservoir as bottom energy is of low order. This gives clue how to segregate the productive reservoir from low productive one. Hence, if proper management is not done in the productive reservoir, it may lead to enormous population growth of weed fishes, which ultimately create a condition most suitable for growth of larger population of catfishes at higher tropic levels. Once these catfishes get a fort hold in the system, the management will be more difficult. It is therefore essential to manipulate the system during the early stage so that vast amount of available energy is properly utilized by stocking with fishes feeding at lower tropic chain. Such phenomenon is witnessed in case of Nagarjuna Sagar and Bhavani Sagar reservoirs. In Nagarjuna Sagar catfishes have established to such an extent that only 20-25 per cent of the available energy is utilized by primary consumers and 75 per cent by Catfishes (Secondary and tertiary).

Pollution is another important problem which considerably affect the healthy biomass production. Some of the reservoirs affected are Gitalsud, Gandhi Sagar, Tungabhadra, Rihand, Bhavani Sagar, Hussain Sagar, Hirakud, Sandynulla, Panchet and Bhairomangala. Eco-degradation of reservoir from thermal power plants located near the reservoir and pollution load from various complex waste water either directly discharged or carried to the reservoir. The complex waste water contains metals like zinc, copper, chromium, cadmium, lead, mercury, pesticides, and coal washeries. Heavy mortality of carps and catfish fingerlings, absence of fish food organisms, plankten, and bottom biota are affected. Besides, dominance of resistant species, other problems *affect* growth rate, reduction in fecundity, high parasitic infection and worms. It is feared that the rapid growing of industries releasing complex wastes may increase such problem in future, if adequate steps are not taken.

It is an established fact that dynamics of spawning and pre-recruitment of carps like meteorological, environmental and biological, but the fact that it requires continued study for a few years to elucidate interferences and interrelationships of these factors for successful spawning success between reservoirs.

Ramchandramenon (1969) from his experience from the eighteen reservoirs of Tamil Nadu options that major carp breeders normally do not breed in main reservoir and prefer migrating up stream for the purpose of breeding. Further, eggs getting washed into the deeper portion are get destroyed. The loss in the maturity condition of fishes, as breeding is rendered impossible in the frequently espoused habitate caused by extreme and almost flood fluctuation (characteristic of D.V.C. reservoir).

On the other hand observations of some successful breeding is also recorded by some workers, Catla, Rohu and mrigal are found to spawn in the tail ends of Tilaiya reservoir while mrigal and calbasa in Panchet hill (Gopalakrishnan and Pal, 1966). Nair (1969) observed large scale escapement of matured fishes from Malampuzha reservoir into the river for breeding during south-west monsoon periods (June-September). Ramchandramenon (1969) observed that in case of Bhavani Sagar reservoir mass occurrence of fertilized eggs of major carps of Bhavani river (6 miles upstream) brought into the reservoir proves that successful repopulation of carps depends on accessibility of the spawning run and in case migrating facility is absent, carps do not propagate. Brood stock ascend about 35 miles upstream. When first flood mingle in the reservoir water mass spawning occurs in shallow sandy river margin. Gopalakrishnan and Pal (1969) observed carps mostly *C. mrigala, C. catla,* bread at the headwater region. The authors observed 104 lakh carp eggs collected in three days. In Panchet although spawning took place at headwater region and in the neighbouring areas of Damodar river, the quantity of spawn was not much due to destruction of breeders by local people. (Spawn collected were 83.2 per cent *C. mrigala, L. bata 10.1 per cent, L. Rohita,* 4.0 per cent and *C. reba* 2.7 per cent only. Spawning ground of major carp has been located at the confluence of Bhadra river with the reservoir (Ghosh and Rao, 1969). It is thus evident from the above findings of various workers that breeding occurs in general in the shallow areas of the head water river or in the confluence of river and reservoir. But in the shallow zone of the main reservoir breeding can not be ruled out. But the quantity of spawn/fingerings which should have shown considerable intensity, is not possible due to (1) large scale capture of brooders that migrate towards the headwaters for breeding. (2) mass occurrence of fertilized eggs which are brought into the deep waters during the receding floods or reservoirs where they can not hatch and are thus destroyed. (3) loss in the maturity condition of fished as breeding is rendered in the frequently exposed habitat due to abrupt fluctuation in the reservoir water and (4) sudden lowering of temperature due to heavy rainfall which affects maturity condition of fishes (Malhotra *et al.,* 1966).

In many cases it has been observed that despite persistent stocking, the transplanted species did not show desired results (proper growth and reproduction). Some of the examples are–*L. fimbriatus,* Common carp, *L. Kontius, P. Carnaticus, C. reba and P. dobius* stocked in Mettur reservoir never recaptured in appreciable quantity. *L. fimbriatus* stocked in Bhavani Sagar did not make any place in the fisheries although it was native to the reservoir. Gangetic carp and *L. fimbriatus* stocked in Vaigai reservoir could not contribute to the fishery. Indian major Carps stocked in Nagarjuna Sagar could not be established as major fishery. Gangetic carp stocked in Malampuzha and Peechi reservoirs could not show substancial increase. Regular stocking of Indian Major carps in Nagarjuna Sagar, Tungabhadra and Krishnaraj Sagar could not contribute to recruitment. Carps stocked in Ravisankar sagar also could not establish due to breeding failure.

From the above, it is clear that mere stocking in reservoirs with spawn/fry does not always mean that spawn/fry will survive and will have a healthy growth, attain maturity and have a natural breeding, while in some reservoirs spawning success is established. These are carps in Ukai and Gandhi Sagar, Catla in Santanur, Calbasu in Bhavani Sagar and Rohu, Catla and Clabasu in Mettur reservoir.

Before stocking, determination of productive potentiality of the reservoir is to be ascertained by the determination of energy flow and nutrient cycle as done by Natarajan and Pathak (1983). In this the basic information on the productivity of reservoir biotopes are discussed. It is only after revealing such basic information; appropriate stocking has to be thought of. For this Jhingran and Natarajan (1969) has done a commendable job in their study on four DVC reservoirs. The important points on which they have stressed for achieving maximum yield are information on self recruitment if any of the stocked economic fishes, actual necessity for stocking and population manipulation through selective stocking or selective elimination and lastly scientific exploitation.

Although diverse opinion are there on the classification of Indian reservoirs, it has been agreed in general to group the reservoirs as smaller (10–∠1000 ha), 1000–∠5000 ha as medium and over 5000 ha as larger reservoirso (Dubey, MS., Sugunan, 1995), Irrigation tanks (mostly seasonal) are also indicated. Available information shows that total number of small reservoirs (∠1000 ha) are maximum in number in Orissa (1433) followed by Rajasthan (389) and Gujarat (115). Irrigation tanks (∠1000 ha) are only lacalised in Tamil Nadu, Karnataka, Andhra Pradesh and Gujarat. Medium size reservoirs (1000–∠5000 ha) are maximum in Andhra Pradesh (32) followed by Rajasthan (30) and Gujarat (28) and one in west Bengal, while larger reservoirs above 5000 ha are maximum in Karnataka (12) followed by Bihar (8) and Andhra Pradesh and Gujarat (7 each) and one only in west Bengal.

From earlier records it is stated that the average yield from 10 states is 18.12 kg/ha/yr. which is the National Fish Production and is based on 422 reservoirs. (291 small, 110 medium and 21 large reservoirs) (Sugunan, 1995). But subsequent information indicates that yield has increased to 49.90 ka/ha in small reservoirs (excluding irrigation tanks), 12.30 kg/ha in medium reservoirs and 11.4 kg/ha in case of large reservoirs (IIMA, 1983).

As stated elsewhere the production can increase considerably in case of medium and large reservoirs like Govind Sagar (16,867 ha), Bhavani Sagar (7876 ha), Ukai (52,000 ha) along with others possibly are not taken into account while projecting the National production level by Sugunan (1995) in all the states. It is thus felt that an upto date survey of data on production of all the states might have enhanced the National production level to a respectable position.

Prospects of pen and cage culture experiments in reservoirs has considerable importance in raising fry/fingerlings to a suitable size before releasing into the reservoir. This process avoids tender fry/fingerlings mortality by directly releasing into the reservoir to a considerable degree. Modification of existing system of pen and cages may be necessary to achieve improved production.

Stress has to be given for standardisation of various nets and gears used in different reservoirs to achieve maximum catch.

Escapement of fish through waste-weirs and canals are considerable. Dubey (MS) has shown that considerable loss of fish takes place from Palwa Dam in Madhya Pradesh. It is reported that during 1975-76, about 7.30 tonnes of fish escaped. This works out to be 36 kg/day. Further, fish escapement through canal is also considerable. Adequate measures as suggested by Dubey (MS) can be thought of. Further, extent of poaching affects nett production of fish from the reservoir. In an estimate it is revealed that the extent of poaching in Rihand reservoir is about 65 t/year, 26 t in case of Kangsabati and 15 t each in Govind Sagar and Hirakud reservoir. In fact poaching is increasing in most of the reservoirs. In smaller reservoir poaching is rather easy as the entire production could have been removed in few nights. Hence, to stop such practice, appropriate action has to be taken.

Thus, fish culture in the reservoir ecosystem are performed in a chain work of stressed environment where stability of water is not possible. High silt precipitation as a result of erosion due to deforestation gradually approaches towards receding the life span of reservoirs with gradual increase in volume development index. Large scale destruction of biomass from productive bays due to their exposure as a result of draw down of water, stocking of fry/fingerlings under the available productive potentiality of the reservoir (monomictic lakes with strong thermal stratification, circulation of water brings locked chemicals of tropholytic zone up to trophogenic zone facilitating fixation and utilisation of energy in peninsular reservoir), biodegradation of reservoir-water and soil from various complex industrial wastes which include heavy metals and pesticides causing destruction of biomass including fish, failure of adequate spawning success due to destruction of mature eggs any fry due to high water level fluctuation, mass killing of ascending breeders towards headwaters, unscientific stocking (where predators and undesirable trash dominate) and management etc. the fish has to complete the life cycle. Besides, global warming which brings changes in the ecological conditions have considerable impal on aquatic life. Fishes grow under such stressed ecosystem to complete their life cycle. Production of such unhealthy fishes which are pooled together to represent them as yield as National Fish production are lately being treated with scientific out look which has shown considerable improvement. All such points are documented and highlighted.

P. Ray

Contents

Chapter 1
Introduction

Damming of streams as stated by Jhingran (1975) has been in vogue since time immemorial. Pre-historic man dammed streams to trap fish. Familiarity with such practice led to the formation of impoundments and diversion of waters for irrigation after the advent of agriculture. With the passage of time and the population influx, the steadily rising demand for water in the country for industrial use agricultural, domestic purposes, power generation, flood control and other water-oriented activities has brought about an accelerated rate of water storage in the form of dams, barrages, weirs, reservoirs etc. to meet the basic need of man.

Monsoon rains demarcate clearly the dry and wet seasons of the year necessitating building of water storage or diversion facilities to tide over the long dry spells, AS such any area having a rainfall of 10 inches (25 cm) or less cannot do without irrigation. Further, in some areas mean annual evaporation in the north exceeds the mean annual rainfall. In the arid zones of Punjab and Rajasthan and with rainfall between 40 to 100 cm, in the drier part of peninsular South India various irrigated methods are adopted to divert river water or store the surface run offs as the case may be. Thus construction of reservoirs large and small by impounding the river run off for irrigational, hydroelectric, water supply, navigation, flood control etc. have become indispensable for proper exploitation of countries natural resources. A list of important reservoirs in different states are indicated in Table 1.

Reservoirs in India (Table 1) as accounted by Jhingran (1975) are located both in the hilly region as well as in the plains and are spread along different river systems (Figure 1). The main reservoirs on the Ganga river system are–Rihand, Matatila, Manaksagar and Sardasagar in Uttar Pradesh; Tilaya, Maithon, Konar, Panchet and Mayurakshi in Bihar; Gandhi Sagar in Madhya Pradesh and Kangsabati in West Bengal. In Brahmaputra river system at present there exists no important reservoir. Gandhisagar is the most important reservoir on river Sutlej of Indus river system; Hirakud is on the Mahanadi of the Mahanadi river system, while Tungabhadra,

Figure 1: River System of India

Nagarjuna Sagar and Nizam Sagar are important in the Krishna river system, and in Kaveri river system–Krishnaraj Sagar, Bhatani Sagar, Mettur Dam (Stanly Reservoir) and Poondi are important reservoirs, m the western coast however, Ukai (Gujarat), Koyna and Shivajinagar and Pawailake (Maharashtra), Malampuzha, Nayyar and Periyar dam (Kerala) are some important reservoirs. Although Godavari river system does not have any major reservoirs, the 188 km stretch of river between Dowlaiswaram and Dummagudem anicuts contribute substantial fishery.

Table 1: List of Important Reservoirs in Different States Jhingran (1975)

State	*Reservoirs*	*River on which Reservoir Situated*	*Location*	*Area (ha) (District)*
Andhra Pradesh	Qsmansagar Dam	Musi	Medan	4,200,980
	Hirnayatsagar Dam	Issi	Medan	3,807,300
	Nagarjuna Sagar Dam	Krishna	Nalagonda	30,303,000
	Mopad Dam	Mammore	Nellore	1,693,860
	Nizamsagar Dam	Mowgina	Nizamabad	14,636,090
	Other reservoirs			78,326,780
				1, 32,968.010
Bihar	Panchet Dam	Damodar	Santal Parganas	7,511 .000
	Maithon Dam	Barakar	–Do–	11,491,830
	Konar Dam	Konar	–Do–	2,792,020
	Tilaya Dam	Barakar	Hazaribag	6,475,000
	Mayurakshi Dam			6,734,000
	Other reservoirs			2,245,530
				37,249,380
Gujarat	Mahi stage II			16,576,000
	Ukai			51,282,000
	Other reservoirs			15,081,570
				82,939,570
Jammu and Kashmir	Lidder Project			116,550
				116,550
Kerala	Periyar Barrage	Periyar	Kbttayam	606,000
	Neyyar Dam	Neyyar	Trivandrum	9,065,000
	Other reservoirs			9, 755, 624
				19,426,624

Contd...

Table 1–Contd...

State	Reservoirs	River on which Reservoir Situated	Location	Area (ha) (District)
Madhya Pradesh	Tawa Multipurpose Project	Tawa	Hoshangabad	29,533,770
	Gandhi Sagar	Chambal	Mandsaur	64,750,000
	Barodia		Shajapur	6,879,040
	Other reservoirs			50,883,930
				1, 52,046,740
Tamil Nadu	Bhavani Sagar	Bhavani	Coimbatore	7,861,840
	Stanley reservoir	Cauvery	Selem	15,343,750
	Poondi reservoir	Kdraliyar	Chinglepet	3,263,400
	Other reservoirs			23,408,63
				49,877,153
Maharashtra	Shivaji Sagar	Koyna	Satara	12,100,480
	Darwa Dam	Darwa	Nasik	3,367,000
	Other reservoirs			1, 35,647,230
				1,51,114,710
Mysore	Tungabhadra	Tungabhadra	Hospet	37,814,000
	Vanivilas Sagar	Vedavathi	Chitradurga	7,252,000
	Linganamaikki	Sharavathi	Shimoga	38, 850, 000
	Krishnaraj Sagar	Kaveri	Mysore	12,924,000
	Other reservoirs			55,932,100
				1,52,772,100
Orissa	Hirakud	Mahanadi	Sambalpur	74,592,000
	Other reservoirs			5,275,830
				79,867,830
Punjab and Himachal Pradesh	Be as Dam	Be as	Kangra	26,418,000
	Govind Sagar	Satluj	–do–	16,838,720
	Other reservoirs			111,370
				43,368,090

Contd...

Table 1–Contd...

State	*Reservoirs*	*River on which Reservoir Situated*	*Location*	*Area (ha) (District)*
Rajasthan	Rana Pratap Sagar			20,720,000
	Bajaj Sagar			12,950,000
	Other reservoirs			10,250,430
				43,920,430
Uttar Pradesh	Sardasagar Dam	Chukasanda	Nainital	7,303,8004
	Nanak Sagar	Desha	–do–	4,662,000
	Matatila			20,720,000
	Rihand	Rend-a tributary	Mirzapur	46,620,000
	Other reservoirs	of son river		46,419,749
				1,25,725,549
West Bengal	Kangsabati			11, 396,000
	D.V.C. Konar Dam			2,331,000
	Other reservoirs			9,840,880
				23,567,880

The fishes contributing to the fisheries of the reservoirs located in the upper reaches of the rivers are three species of mahseers, namely *Tor putitora, T. mosal* and *T. tor; the Katli, Acrossocheilus hexagonolepis* the snow trout, *schizothorax plagiostomus;* medium carps, *labeo dero* and *L. pangusia* and the goonch, *Bagarius bagarius.*

The reservoirs located in the middle and lower reaches of the rivers harbour the Indian major Carps, *Catla caffa Labeo rohita Cirrhina Mrigal* and *Labeo Kalbasu,* other carps *viz. Labeo gonius, L. bata, L. boga, L. Boggut, Pontius sarana, Chagunius chagunio* etc. The major catfishers contisbutug to the fisheries of the reservoirs of this region are *wallango attu, Silonia silondia, Pangasius pangasius, Rita, rita, Mystus aor, M. Seenghala.* etc. Smaller Catfishes represented in these reservoirs are *clupisoma garua, Eutropiichthys vacha, Mystus cavasius,Onpok bimaculatus* etc.

In the Brahmaputra river system there does not exist any important reservoir.

In Punjab, Govindsagar on river Sutlej is the most important reservoir with developed fisheries. The fish species are commonly those found in the upper reaches of the Gangetic system. *Tor putitora* and *Labeo dero* were dominant throughout the reservoir. Besides *Schizothorax Plagiostomus, Crossocheilus latius, Garra gotyla, clupisoma montana, Mastocembelus armatus* and *Mystus bleekeri* occurred largely in upper sections and *Cyprinus carpio var, specularis, Mystus seenghala and Labeo bata* were also found in the middle and lower parts of the reservoir.

The fish fauna of Hirakud in the Mahanadi river system resembles that of the Ganga river system. Besides the Gangetic carps and cat fishes *Tor mosal* var. *mahanadicus, Labeo finbriatus* and *Rita Chrysea* are some of the commercially important species.

In the Godavari river system there is no dam on the river Godavari. The impounded stretch of 188 km between Dowlaiswaram and Dummagudem anicuts is lacustrine in character and the commercially important species of the impounded area are *Cirrhinus mrigala, Labeo fimbriatus, L. Calbasu, L. rohita, Cirrhinus horai, Catla Catla, Mystus seenghala, M.aor. Bagarius bagarius, Pangasius pangasius* and *Silonia childreni.* The prawn, *Macrobrachium malcolmsonii* form the major fishery.

Tungabhadra, Nagarjuna Sagar and Nizam Sagar are the important reservoirs of the Krishna river system. The important endemic species are: *Labeo fimbriatus, L. Calbasu, L. porcellus, L. potail, L. bata, L. boga, L. boggut, puntius Kolus, P. pulchellus, P. dobsoni, P. curmuca. P. sarana, Tor Khudree, T. mussulah, Cirrhinus reba, C, fulungee, Osteobrama vigorsil, O. belangeri, Thynnichthys sandkhol, Wallago attu, Mystus aor, M. Seenghala, M. punctatus, M. Cavasius, Silonia childreni, Eutropiichthy goongwaree, Pseudeutropius taakree, Rita pavimentata, Ompok bimaculatus, Bagarius bagarius* etc. *Puntius Kblus, P. pulchellus* and *Thynnichthys sandkhol* dominates in the Tungabhadra reservoir, Anjanpur reservoir and Nizam Sagar respectively. Indian major carps introduced in many reservoirs of the Krishna river system have established themselves. *Labeo gonius* and *L. pangusia* introduced along with the Indian major Carps also occur. *Tor neilli,* a species of mahaseer is endemic to Tungabhadra and Krishna rivers.

The more important reservoirs in the Cauvery river system are Krishnarajasagar in Mysore and Bhavani Sagar, Mettur Dam (Stanley reservoir) and Poondi Reservoir in Tamil Nadu. The dominant species of this system are: *Puntius dubius, P. Carnaticus, P. sarana, P. micropogon, Cirrhinus Cirrhosa, C. reba, Labeo Kontius, L. Calbasu, L. fimbriatus, L. potil, L. porcellus. Tor Khudree, I. mussullah. Aerossochelus, hexagonolepis, Bagarius bagarius, Wallago attu, Mystus punctatus. Silonia childreni, Pangasius pangasius.* In addition to the above *Catla catla, Labeo rohita, Cirrhinus mrigala, Cyprinus carpio, Osphronemus goramy* and *Etroplus suratensis* have been stocked in many of the reservoirs of this river system where they have established.

In reservoirs situated on other rivers, originating from the western Ghat and flowing into the Arabian Sea, the important Peninsular economic species are: *Labeo nigrescens, Puntius Thomassi, P. jerdoni,P. filamentosus, P. curmuca, P. pulchellus, Qnpok malabaricus* etc. In Malamphuza reservoir, *Catla catla, Labeo fimbriatus, Cirrhinus Cirrhosa, Tilapia mossarnbica* and the minor carp *C. Carpio var. specularis* have been transplanted.

Besides the species mentioned under the reservoirs of different river systems there are several others which occur occassionally in different reservoirs. These are *Channa spp., Mastocenibelus* spp., *Notopterus* spp., *Aspidoparia, Qxygaster, Rasbora amblypharyngodon, Osteobramg* and small scaled barbells, *Pontius* spp.

Chapter 2
Background History

The river valley development project since India's independence are to serve the irrigation needs of agriculture, power generation, flood control and water supply for domestic and industrial use and the dam design accordingly varies from reservoir to reservoir. The utilization of this resource for fish development, however, took a low priority and this is in a way reflected by poor fish yield which, by all accounts available, may not exceed a national average of 10 kg/ha. The reasons for this low level of fish yield from reservoirs are not far to seek and may be attributed to inadequate understanding of the ecosystem, fish behaviour and fish stock management. The research and management gaps were sought to be bridged through an All India Coordinated Research Project on Ecology and Fisheries of Freshwater Reservoir which was launched in 1971 with well defined objectives which inter alia included studies on trophic structure and function of the ecosystem, material (nutrient) recycling, physico-chemical features, ecological production functions, fish behaviour, recruitment and other population parameters and development of management principles with a view to achieve optimum fish productivity and fish yield. These studies were carried out by CIFRI and other state-level research organisations on man made impoundments outside the field of the Project have resulted in a wealth of information on reservoir ecosystem and have led to the development of sound management principles for reservoir fisheries development which has relevance not only to India but to tropical impoundments in general in South and South-East Asia.

Fairly a large number of studies have been made all over the world regarding management practices of reservoirs with a view to increase yield per ha. About 3 major symposia (1960, 1969 and 1973), dealt with the world problems of large man-made lakes outlined guidelines for policy and planning for reservoir fishery management. Subsequently a symposium on utilisation of Inland Fishery Resources' particularly on reservoirs was organised by Indo-Pacific Fisheries Council held in

1976, which was emphasised on collection of more data on different aspects of reservoir management. In India attention for development of Fisheries in reservoir commenced in the beginning of First Five Year Plan, and since then number of symposia were organised by Central Inland Fisheries Research Institute/Central Inland Fisheries Education in the year 1969 and 1979. A coordinated project on 'Ecology and Fisheries in Reservoirs' was also initiated by Central Inland Fisheries Research Institute in 1971 which yielded encouraging results (Dubey, MS).

The total surface area of reservoirs in India has been estimated at 3.00 million ha and it is expected to increase 100 per cant *i.e.* 6 million ha by the turn of the century (NCA, 1976) According to Bhukaswan (1978) the area of the reservoirs in India is 20,000 km^2.

The reservoir resources have been grouped into three main categories (IIMA, 1983).

Large reservoirs: With more than 5000 ha in area.

Medium reservoirs: Between 1000–∠5000 ha.

Small reservoirs: Below 1000 ha

According to the latest report (IIMA, 1983) there are 975 reservoirs in the country. Out of this, there are about 550 major reservoirs. Of this, 5 per cent are very large (above 10,000 ha.), 3 per cent are large (5,000 to ∠10,000 ha), 26 per cent medium (1,000 to ∠5,000 ha) and rest are small. Thus the number of medium and small reservoirs comprise of 92 per cent. Some of the large reservoirs in the country are:

20,000–Above at FRL

Hirakud: 71,000 (Orissa)

Gandhi Sagar: 66,000. (Himachal Pradesh)

Ukai: 52,000 (Gujarat).

Rihand: 46,538 (Uttar Pradesh)

Tungabhadra: 37,814 (Karnataka)

Nagarjuna Sagar: 28,000 (Andhra Pradesh)

Pongal Reservoir: 24,627 (HP)

10,000–∠20,000 ha at FRL

Ponam: 18740 (Gujarat)

Govind Sagar: 16867 (HP)

Krishnaraj Sagar: 13200 (Karnataka)

Nizam Sagar: 12,800 (AP)

Maithan: 11,496 (Bihar).

Chapter 3

Physiographical and Meteorological Conditions

Physiographical and meteorological conditions are vital aspects normally considered at the time of dam and barrage used for various purposes. Chopra *et al.* (1949) has given a brief account of the above conditions in his valuable document "Poisonous plants of India".

Physiographical Conditions

The physical features of the country also present a variety probably unrivalled by any other area of similar size on the globe, and next to its climate exercise the most important influence in determining its vegetations. In the North, extending from Assam to Kashmir and running in a General North-westernly direction lies the loftiest and most massive range of mountains in the world, the Himalayas, with one of its peaks Mount Everest, rising to 29,002 ft. above the sea level. This is separated from the Peninsular India properly by the Indo-Gangetic depression, a broad belt of the country through which in the west, the Indus and its tributaries drain towards the South westernly direction to the Arabian sea, which is not far away from where it has originated. The Ganga takes its birth to drain the great fertile plains of the North-East India comprising Uttar Pradesh, Bihar and West Bengal.

India, South of the Indo-Gangetic depression is characterised by a backbone of mountains in the west, known as western ghats, which run more or less parallel to and at no great distance from the coast of the Arabian sea. On the north by the Aravali range and the rising uplands of the Central India and Chhotanagpur and on the east by the range known as Eastern Ghats. Roughly, these bounds and contain the great central Plateau, which is itself intersected mainly from the Westernly to an easternly direction by the western system of the Godavari and Krishna and the minor streams draining into the Bay of Bengal. To the south of the peninsular, the ghats taper off in the Nilgiri and Cardamon Hills to gentle smooth rounded slopes of green uplands. Between the ghats and the sea there is a comparatively narrow flat or gently sloping belt. (Chopra *et al.*, 1949).

Table 2: Rainfall and Evaporation at Selected Stations in the Ganga Basin

Station	*Mean Annual Rainfall (mm) (1931-1960)*	*Mean Annual Evaporation (mm) (1959-1975)*	*Difference Between Annual Rainfall and Evaporation (mm)*
Delhi			
Safdarganj	714,2	2223.8	–1409.1
Punjab			
Hissar	389.9	2579.1	–2189.2
Uttar Pradesh			
Agra	781.6	2232.2	–1450.6
Allahabad	1027.3	1843.6	–816.3
Bareilly	1068.3	1908.7	–840.4
Gonda	1294.0	1803.5	–509.5
Bihar			
Gaya	1097.3	2220.1	–1122.8
Patna	1109.6	1886.1	–776.5
West Bengal			
Assansol	1391.8	1456.1	–64.3
Rajasthan			
Kota	841.6	2591.4	–1749.8

Source: Das Gupta, S.P. (1984).

Mateorological Conditions

India presents a great variety of mateorological conditions and features than perhaps any other area of similar size in the world, and furnishes the typical large scale example of the alterations of seasons known as the monsoon. It may be rightly described as land of diversity. During the winter third of the year, the general flow of the surface air strata is formed land to sea and then over the Indian seas as a north-east monsoon. It is a season of winds of continental origin and great dryness. The summer third of the year brings a complete reversal of this condition in a flow from sea to land of the moist winds of the south-west monsoon which consequently is a season of much humidity, clouds and frequent rain. Between these principal seasons of the year are the transitional periods of the hot weather months, April and May and the retreating south-west monsoon, October and November. Rainfall with which is associated the humidity of the atmosphere, and which more than any other mateorological condition exercises an influence on the vigetation, varies from about 430" inches at Cherapunji in the hills of Assam to the north-east of Bay of Bengal and from about 350" in certain portions of the Western Ghat to as little as 5" or less per

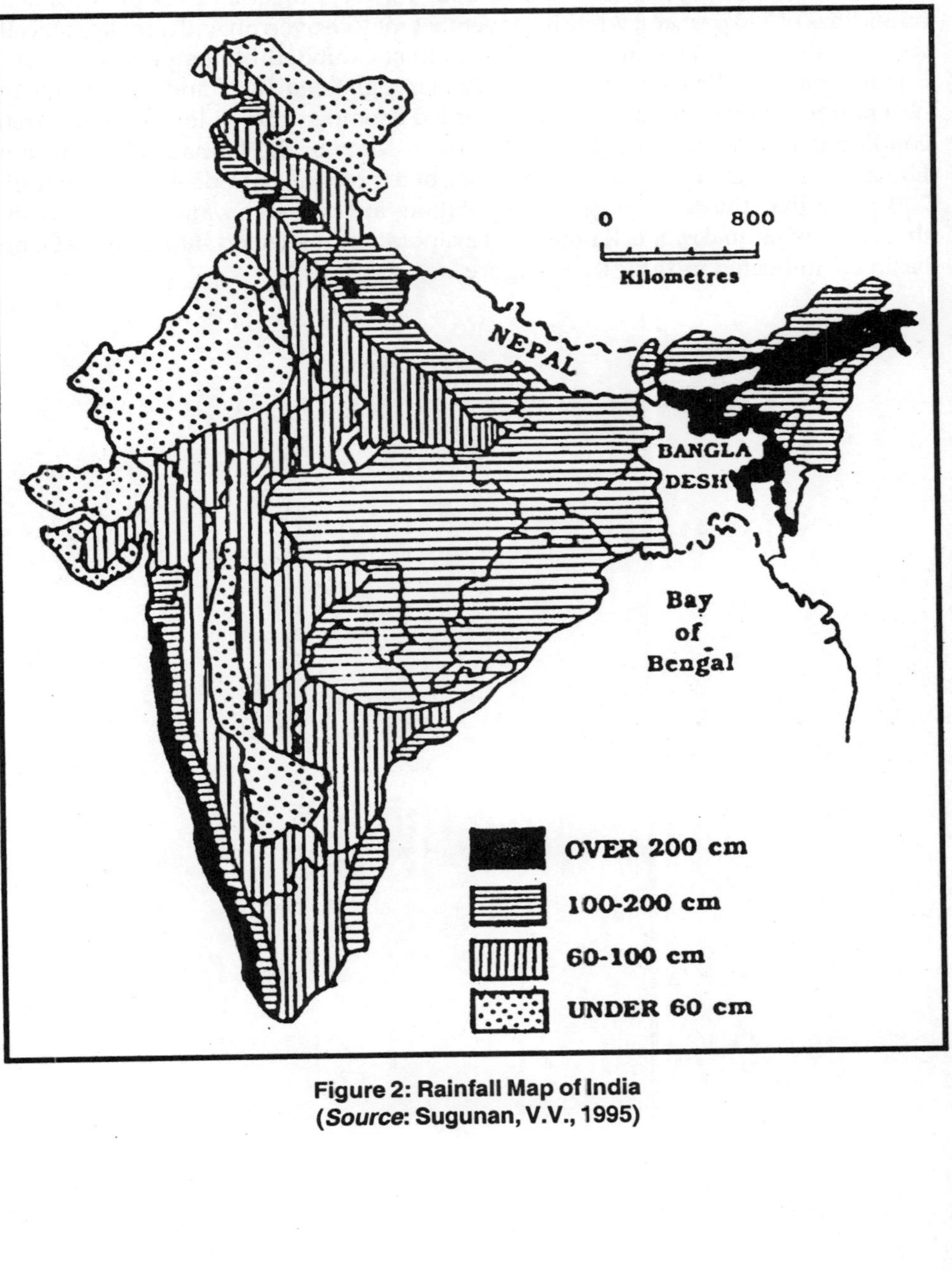

Figure 2: Rainfall Map of India
(*Source*: Sugunan, V.V., 1995)

annum in the desert areas of Rajasthan in the north-west. During the south-west monsoon from June to September the air is supersaturated with moisture in the coastal districts and in the hills, while during the dry weather, it may in certain districts be so dry as to defy the ordinary methods of calculations of humidity. The hill stations in the Himalayas may be shrouded in clouds for days together in September with humidities of 100 per cent, while in November may be overrun with air of practically zero-humidity. No less striking is the contrast exhibited by temperature; that at Sriganganagar in Rajasthan has several times registered 122°F and over, while the observatory at Dras in Kashmir has recorded a temperature as low as–49°F. Arctic conditions prevailed throughout the year in some of the Himalayan and trans-Himalayan ranges. The mean annual range of temperature at Cochin in south India, 20°F is less than the daily range at many stations in North India, and only about one-third of their annual range. Rainfall and evaporation at selected station in the Ganga basin are indicated in Table 2 and Figure 2.

Chapter 4

Forest Cover and Soil Erosion

Man has started over-exploitation and denudation of natural resources especially soil and water to meet their requirements. The need for food, fuel and water has increased tremendously as a result of continuous growth of both human and animal population. This is liable lead the harming immediate surrounding and threaten the suitability and viability of their environment. In the past few decades, tremendous damage to soil, forests, grasslands and many other forms of plant life has occurred. The report of the National Committee on environmental planning and coordination (Bhardwaj and Singh, 1985) points out that only 12 per cent of the country's land surface is actually under adequate protective tree-cover as against the largest of 33 per cent prescribed by the National Forest Policy formulated in 1952. If soil and vegetation are not in balance, as often they are not when influenced by poorly managed human activities, erosion is accelerated which may result, disastrous consequences. Their management and over-exploitation of soil-water-plant system by growing population is the root cause of environmental degradation.

It is alarming to see the vanishing forests in different parts of the country of which West Bengal is leading *i.e.* 324,000 ha (*The Statesman*, June 29, 1989). This was followed by Karnataka, Uttar Pradesh, Maharashtra, Andhra Pradesh, Kerala and Madhya Pradesh, Gujarat, Jammu and Kashmir, Rajasthan, Assam, Bihar and Tamil Nadu (Figure 3). The consequences of deforestation are too well known to need detailed forestation. Such deforestation leads to a tailspin of soil erosion, siltation of water courses and reservoirs, lack of percolation of water underground resulting high flood. (Tables 4, 5, 6, 7, 8).

Dhruva Narayan and Rambabu (1983) reported a total of 5334 million tonnes of soil is being annually eroded. The country's river carry an approximate quantity of 2,052 m tonnes, nearly 480 m tonnes are deposited in various reservoirs, and 1572 m tonnes are washed into the sea every year

Figure 3: The Vanishing Forests of India
(The Statesman, Kolkata, June 29, 1989)

According to the latest Forest survey of India (using satellite imagery) the forest cover in the country is reduced by 1907 sq.km. in 4 yrs. (*The Telegraph*, 1990). It is further reported that the assessment made in 1987 (based on imagery of 1981-83) the country had a forest cover by 6,42,041 sq.km. which during 1989 (based on imagery of 1985-87) reduced to 6,40,134 sq.km.

Although some states showed a loss in their forest cover, some gained in their area due to the conservation measures taken. These are indicated in Table 3.

Table 3: Showing Loss/Gain of Forest Cover in Some States

Sl.No.	*State*	*Loss of Forest Cover Area (sq.km.)*	*State*	*Gain in Forest Cover Area (sq.km)*
1.	Orissa	6,026	Arunachal Pradesh	8,263
2.	Maharashtra	3,358	Madhya Pradesh	5,442
3.	Andhra Pradesh	2,283	Uttar Pradesh	2,401
4.	Bihar	1,814	Goa	Area not available
5.	Assam	Area not available	Himachal Pradesh	
6.	Gujarat	–	Manipur	–
7.	Haryana	–	Nagaland	–
8.	Jammu and Kashmir	–	Punjab	–
9.	Karnataka	–	Rajasthan	–
10.	Kerala	–	Sikkim	–
11.	Mizorarn	–	Andaman and Nicobar Island	–
12.	Meghalaya	–	Chandigarh	–
13.	Tamil Nadu	–	Delhi	–
14.	Tripura	–		
15.	West Bengal	–		
16.	Union Territory of Dadra and Nagar Haveli	–		

Source: *The Telegraph*, May, 29, 1990.

Chapter 5
Deforestation vis-à-vis Soil Erosion

Samraj *et al.* (1985) conducted some experimental work to show the difference in the rate of silt loss between well and ill managed water sheds of different categories in Nilgiri. This will undoubtedly through, light on the importance of forest cover. The results pertaining to run off, soil loss and peak floods obtained under different land uses are presented in Table 4.

Table 4: Run Off, Soil Loss and Peak Floods in Nilgiri Water Sheds (Bhimani *et al.*, 1983)

Particulars	*Run off as P.C. of rainfall*	*Peak Floods*	*Soil loss t/ha/annum*
(i) Forest			
Well managed	25–43	low	0.00–0.01
ill managed	10–30	High	20–60
(ii) Grass land			
Well managed	10–30	Low	0.00–0.01
ill Managed	10–30	High	20–60
(iii) Agriculture Land:			
Well managed	6–30	Medium	0.5–4.0
ill managed	10–30	High or very high	20–60
(iv) Urban Land:			
Well managed	10–20	Medium	3–10
ill managed	15–30	Very high	60–120

Source: Samraj *et al.* (1985).

The above study highlights the importance in favour of perennial vegetation consisting of trees, shrubs, herbs and grass for conserving the soil effectively, regulate water in mountain streams and to protect dams against siltation. The rate of siltation in reservoirs goes higher than expected. Bihar recorded 214, Assam 134 and Uttar Pradesh 99 major floods during 12 years' period. Table 5 indicates rate of siltation in some major reservoirs in the country as obtained from sedimentation surveys conducted by the Central Water Commission. It is evident that absence of adequate flow from reservoirs causes silting in rivers which ultimately helps flooding. The sedimentation surveys show that the actual incidence of sedimentation per 100 km^2 catchments areas in some cases are several times more than the assessed estimation (Tables 5, 6, 7) in course of few year's time.

Loss of storage capacity of reservoirs due to silting is by far one of the most serious problems created by soil erosion. All these indicate that the reservoir is fast loosing its life span and it will loose all its beneficial qualities one after another, whose impact will be immense.

Chapter 6
The Soil

Soil plays an important role in determining the fertility of fish ponds and reservoirs. The characteristics of different soils and their availability as outlined by Jhingran (1975) are discussed below. The fertility of soil refers to nutrients releasing properties for the benefit of aquatic crop (Tables 3, 4 and 5).

Alluvial Soil

It covers an area of about 1,500,000 sq.km. in the Indo-Gangetic plain, Punjab, Gujarat, Orissa, Tamil Nadu, Kerala etc. and is formed by silt deposited by numerous river system. Alluvial soils, though deficient in nitrogen but contains adequate quantities of potash, soda and phosphoric acid. Their porosity, texture etc. tend to provide for good drainage.

It is thus clear from the above tables that actual silt precipitation is several times more than what was assumed (Table 6). As a result the reservoirs in the long run will fail to serve the purpose for which they were constructed. The tables show that during construction of reservoirs, proper attention was not given towards quantum of silt erosion that can take place from the catchment area. Based on such silt precipitation, Bera (1984) has calculated the life span of some reservoirs indicated below in Table 7.

Red Soil

Red soil are common in Tamil Nadu, Mysore, south-east, of Maharashtra, east of andhra Pradesh, Madhya Pradesh, Orissa and Chhotta Nagpur covering a total area of about 3,50,000 sq.km. Its area extends into Santal Parganas, Birbhum district of west Bengal, Mirzapur, Jhansi and Hamirpur districts of Uttar Pradesh. These soils are generally poor in nitrogen, phosphorus, lime, iron oxide and humus with predominance of the mineral kaoline (hydrated aluminium silicate). Red soil possess little capacity for absorption of moisture. Their capacity for retention of bases (potassium, calcium and magnesium) by exchange is also limited. Phosphorus gets more readily fixed in these soils, the capacity for which increase with decrease in pH.

Table 5: Sediment Data of Selected Reservoirs

Reservoir	Year of Impounding	Net Catchment Area in km.²	Capacity in M hours		Total Sediment Deposited		Annual Av. Rate of Siliting in ha.m.	Annual Rate of Silting ha.m. 100 km.²		Corrected After Conversion of Suspended Detail	Remarks
			Gross	Dead	ha.m.	Period upto		Assumed	Observed		
Bhakra Nangal	1958	56980	0.987	0.243	77670	1981	–	4.29	5.93	–	Capacity surveys were being repeated every year from 1965-77 and there after every year.
Hirakud	1956	83395	0.810	0.226	147859	1982	–	2.52	6.82	-	Inflow and outflow method.
Maithan	1956	5206	0.136	0.002	10962	1971	685	1.62	13.10	13.10	Capacity surveys done in 1965 and 1970.
Panchet	1956	9816	0.150	0.018	17663	1974	978	2.47	10.00	10.00	Capacity surveys done in 1965 and 1970
Mayurakshi*	1955	1792	0.061	0.007	4710	1970	294	3.61	16.43	16.43	Capacity surveys dam in 1965 and 70. Sediment in flow data.
Ramganga	1975	3134	0.245	0.025	10259	1981	–	4.27	54.56	–	
Kangsabati	1965	–	0.114	0.015	862	1970	142	3.27	3.76	9.65	–Do–
Ghod	1966	3029	0.022	0.007	2765	1970	553	3.61	15.24	15.24	Inflow outflow method
Dhantiwar	1965	2862	0.046	0.008	646	1968	130	3.61	4.52	3.95	–Do–

Contd...

Table 5–Contd...

Reservoir	Year of Impounding	Net Catchment Area in km.²	Capacity in M hours		Total Sediment Deposited		Annual Av. Rate of Siliting in ha.m.	Annual Rate of Silting ha.m. 100 km.²		Corrected After Conversion of Suspended Detail	Remarks
			Gross	Dead	ha.m.	Period upto		Assumed	Observed		
Beas Unit II	1974	12561	0.0858	0.129	23152	1982	–	4.29	23.09	-	Sediment inflow data
Ukai	1972	62224	0.851	0.142	49600	1971	–	1.37	11.39	-	Inflow outflow method
Tawa**	1974	5983	0.231	0.022	5300	1973	670	3.61	11.51	15.60	Sediment inflow data
Nizam Sagar[a]	1967							0.29	6.57		
Tungabhadra[b]	1953							4.29	6.11		
Shivaji Sagar[c]								3.24	15.24		
Gandhi Sagar[d]	1960							3.61	10.05		
Govind Sagar	1963	56980						4.29	6.00		

a,b,c,d: Data provided by Bharawaj, S. P. and Ravinder Singh, (1985); e: (Joshi 1990).

Source: Prem Narayan, Kathuria, o. P. and Bathla, H. V. (1985).

* The sediment rate for the period 1965-70 (5 years) is 22.77 ha.m/100 sq.km./Year

** Silt data pertain to the period from 1958-70.

Table 6: Quantum of Annual Silting in Some Reservoirs (acre ft.)

	Name of Reservoir	*Assumed Annual Silt Load Precipitation (acre ft.)*	*Actual Annual Silt Load Precipitation (acre/ft.)*	*Percentage of Increase of Silt Load*
1957	Maitan (11491 ha)	684	5980	774
1957	Panchet (7511 ha)	1982	9553	381
1955	Mayurakshi (6734 ha)	538	2000	271
1963	Bhakra Nangal	23000	33475	45.5
1975	Ramganga	1098	4376	298.5
1953	Tungabhadra (37814)	9796	41058	319.1
1967	Nizam Sagar (12800)	530	8725	154.6
1971	Ukai (52000)	7448	21758	192.1

Source: Bera (1984). Based on 1972 Irrigation Commission Report.

Table 7: Life Span of Some Reservoirs (Bera, 1984)

Reservoirs	*Calculated Life Span Based on Assumed Silt Precipitation (yrs.)*	*Calculated Life Span Based on Actual Silt Precipitation (yrs.)*
Maithon	212	153
Panchet	216	138
Mayurakshi	872	200
Bhakra Nangal	403	291
Hirakud	386	147
Gandhi Sagar	930	350
Tungabhadra	311	245

Black Soils

Black soils are highly argillaceous, made up of fine grains dark in colour with a high proportion of calcium and magnesium carbonates, iron, potassium, magnesium and aluminium, but poor in phosphorus, nitro gen and organic matter and are available in Maharashtra, Western part of Madhya Pradesh, parts of Madhya Pradesh, parts of Mysore, Andhra Pradesh, Gujarat and some parts of Tamil Nadu including the districts of Ramnad and Tiruneveli to the extreme South. Black soil is highly impervious to water and becomes sticky when wet,

Laterite and Lateritic

Laterite and lateritic soils are compact and composed of vesicular rocks having a mixture of hydrated oxides of aluminium and iron with small amount of manganese oxides and titanium. These soils occupy a total area of about 248,000 km^{-2}. Laterite

Table 8: Statewise Distribution of Major Floods in Indian Rivers During 1969-1981

Year	Andhra Pradesh	Arunachal Pradesh	Assam	Bihar	Gujarat	Madhya Pradesh	Maharashtra	Orissa	Uttar Pradesh	West Bengal	Delhi	Rajasthan
1969	–	–	–	–	–	1	1	–	2	–	–	–
1970	–	–	5	3	3	3	–	–	2	1	–	–
1971	–	–	–	29	1	–	–	5	11	9	2	–
1972	–	–	5	2	3	4	–	1	–	–	–	–
1973	–	–	4	13	8	8	–	7	13	7	–	–
1974	–	–	11	20	2	2	–	2	10	–	–	–
1975	3	–	4	24	3	4	–	2	11	–	1	–
1976	4	–	7	12	2	–	–	2	17	1	–	–
1977	–	–	10	8	4	–	–	1	2	2	–	–
1978	–	–	12	13	2	–	–	–	19	10	1	–
1979	–	–	14	12	1	–	–	–	2	1	–	–
1980	–	–	41	41	2	2	–	1	17	7	–	–
1981	2	1	2	18	1	–	–	–	3	2	–	1

Source: Prem Narain, O. P. Kathuria and H.V.L. Bathla (1985).

soils are poor in phosphorus, potassium, calcium and nitrogen with pH ranging from 4.8–5.5 and low base-exchange capacity. The varietal distribution of lateritic soils are in Bihar, West Bengal, Orissa, Maharashtra, Mysore and Tamil Nadu.

Forest Soil

The forest soils comprise about 285,000 sq.km. in India and are very rich in organic matters derived from forest growths. These are of two types:

1. Soils formed under acidic conditions with the presence of acidic humus and low base status, and
2. Soils formed under slightly acidic or neutral condition with a high base Status.

The Desert Soil

Cover about 142,000 sq.km. of arid and semiarid areas in Rajasthan and South Punjab lying between the Indus and Aravallis. These are derived mostly from brown sand. Some of them contain high percentage of soluble salts, are alkaline and poor in organic matter.

Saline Soils

Saline soils abound in Sodium Chloride and Sodium Sulphate and the alkaline soils in Sodium Carbonate. These soils have an alkaline action and it is the carbonate that renders the soil sterile. They are highly impervious. Bacterial activity and organic matter are practically absent in such soils due to lack of vegetation.

Peat Soils

Peat Soil are found in Kuttanand (Kerala), in coastal areas of Orissa, in Sunderbans and other places of west Bengal; in the Central portion of North Bihar in the district of Almora (Uttar Pradesh) and in the South-east of Chennai. Peat soil areas have an accumulation of large quantities of water soluble alkaline soils. The water logged areas formed by dried river basin and lakes in alluvial and coastal areas are generally blue or blue-black in colour due to the presence of iron (ferrous) and also contain varying amounts of organic matter. Soil fertility maps of India showing their distribution are indicate in Figures 4, 5, 6.

Table 9: Annual Loss of Potassium and Nitrates (N) in Tonnes Through Some River System

River System	*Potassium*	*Nitrate (N)*
Godavari	22,197	5,410
Sabarmati	1,589	193
Narmada	8,509	3,206
Krishna	7,312	1,470

Source: Hydrology Review, Vol. 3, (1977); In: Bhardwaj, S.P. and Revender Singh (1985).

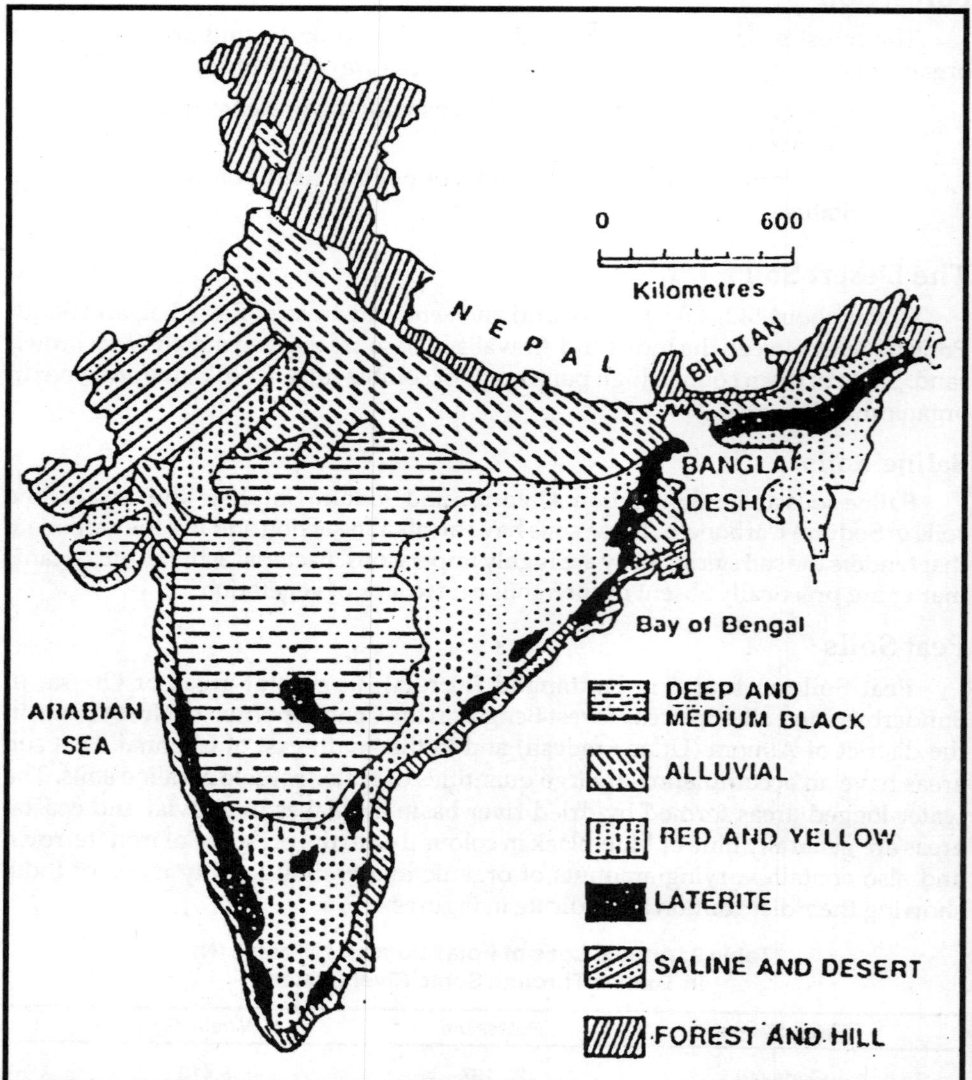

Figure 4: Soil Map of India
(*Source*: Sugunan, V.V., 1995)

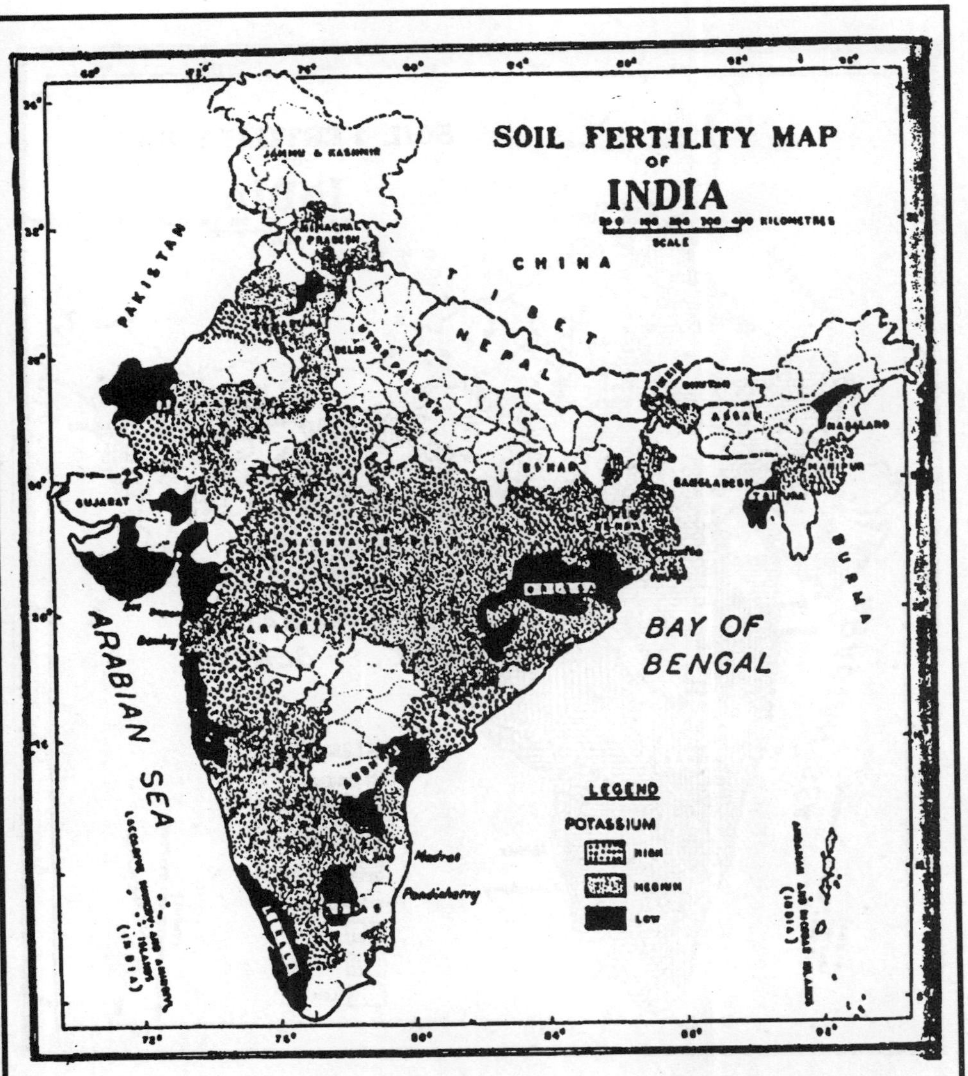

Figure 5: Soil Fertility Map of India Showing Distribution of Potassium
(*Source*: Jhingran, V.G., 1975)

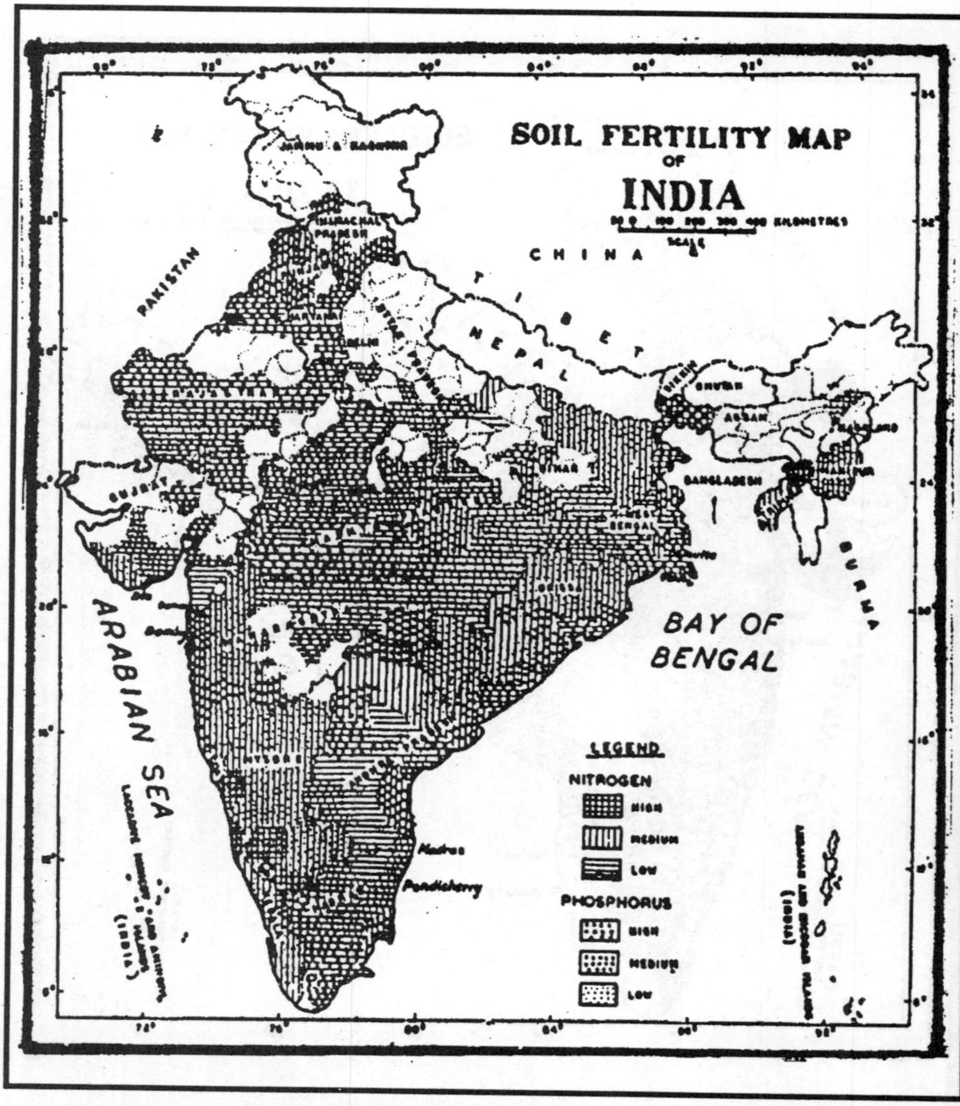

Figure 6: Soil Fertility Map of India Showing Distribution of Nitrogen and Phosphorus (*Source*: Jhingran, V.G., 1975)

Nutrient Loss and Soil Fertility

It has been estimated that eroded top soil carries away about 2.5, 3.3 and 2.5 m tonnes of nitrogen. Phosphorus and Potash respectively and huge quantities of other essential nutrients every year (Swaminathan, 1973). The loss of N.P.K. is more than their production. Erosion removes nutrients as solution and as physical removal of soil and organic matter Figures 3, 4 and 5. Tamhane *et al.* (1959) reported that soil loss in runoff on 3.5 percent slope was much more clayed and richer in organic-carbon, nitrogen and potash as compared to the original soil. The eroded soil particles carry attached nutrients from fields into stream, ocean and lakes. Finer portion of the soil is always removed first and this fraction is richest in fertility. An idea of the quantum of such loss through some peninsular river discharges which is quite considerable are indicated in Table 9.

Chapter 7

Sequence of Eco-Morphological Events Following Reservoir Formation

Soon after impoundment there is a change in the water ecosystem from lotic to lentic and there occurs a phase of high initial fertility eaused by nutrients leaching from the hetherto unflooded substrate, submerged forests and other organic matter. With the change in the ecosystem the biomass is also changed. This accelerates the growth of bacteria, plytoplankton, zooplankton and benthos. In this new habitat which has greatly reduced fish population per unit volume, zooplankton forms are under little predation pressure and build up large populations dependent 'on preponderance of plantonic algae. With the larger forms of zooplankton available for food, the general picture is one of good growth of fishes and good fishing success. The geographic location affects the metabolism of a reservoir through nutrient supply, shape of basin and efficiency with which the climatic factors able to act in the dynamic exchange. In addition, the hydrological cycle (inflow and outflow) also play important role. They all have varying effects on final productivity.

The edaphic factors affect the supply of dissolved nutrients in the reservoir water. The extent of drainage area, its rate of erosion and run off are important in limiting the supply of nutrients to the reservoir. Soil erosion influences the reservoir productivity to a great extent. Area, mean depth and regularity of shore line have the most significant bearing on productivity of a reservoir. The notable biotic and abiotic factors that may influence the productivity at various tropic levels are listed in Table 10 as positive and negative factors.

Table 10: Abiotic and Biotic Factors Affecting Productivity of Reservoirs at Various Trophic Levels

Sl.No.	*Positive/Augmentative Factors*	*Major Effects Unknown*	*Negative/Reductive Factors*
1.	High shore line development coves, bays etc.	Sedimentation of inorganic materials	Low transparency in floods due to inorganic turbidity
2.	Low mean depth (Less than 18 m)	High rate of evaporation	High mean depth
3.	Existence and extent of marginal vegetation	Contributions of auto autochthonous nutrients	Erosion in the reservoir watershed area
4.	Optimum nutrient levels	High surface temperature during summer (In North India)	Reduction of quantity of water flowing into reservoir
5.	Nutrient enrichment during floods	Low water temperature during winter (In North India)	Large water level fluctuations creating large aridal (barren littoral
6.	Moderate to long growing season	Community interrelationship	Low level of dissolved oxygen in parts of hypolimnion
7.	High frequency of phytoplantium blooms	-	Pollutions in the reservoir watershed areas
8.	Moderate macrophytes	-	Plytoplarikton biomass mainly blue greens
9.	Periphyton density increasing markedly	-	Relative low fish species diversity indicating low stability and a potentially low resilience to stresses
10.	Well established plankton and benthos	-	Imbalance fish populations favouring predatory trash species
11.	Tree and bush clearing	-	Low abundance and diversity of terrestrial vegetation, hence early successional stage
12.	Conditions permitting passage of migratory fish and their economic worth, if to be created	-	Relatively low environmental heterogeneity
13.	Introduction of fish adopted to lentic conditions	-	Low diversity of plankton' and benthos
14.	Utilization of modern fishing gear and optimization of fishing effort	-	Low diversity of aquatic macrophytes
15.	Enforcement of fishery regulations	-	Exposure of nests of Catfish, *Aorichthys seenghala* due to severe draw downs

Source: Arun G. Jhingran, 1986.

Chapter 8

Silt Load Affecting Morphological Character of Reservoir

Generally reservoirs are constructed in the hilly gorges of deeply infested forest areas, with the population influx deforestation becomes essential for their habitation. As a result every year large quantities of loose top soil alongwith other organics are brought into the reservoir alongwith the monsoon water. Such situation causes reduction of dead storage, live storage and finally the total storage. Thus morphoecological changes take place in the reservoir which becomes more acute during drought or late monsoon. The water holding capacity of the reservoir is thus lost year after year indicated by different workers (Tables 5, 6, 7) and their impact on aquatic life is likely to be affected.

Chapter 9
Rheology of Reservoirs

Information on morphometry especially shore development and volume development index may have significance in understanding some aspects of reservoir productivity. This gives information on the degree of regularity or irregularity of the shore line and nature of reservoir bed. If the value of shore development index is greater than unity, shore development index of the reservoir is irregular, while in the case of volume development if the value is less than unity the reservoir basin wall is convex towards the water and when the value is greater than unity the basin walls are concave. Available information shows that quite a good number of reservoirs, large and small are having highly irregular shore line which indicates presence of shallow littoral zones which are supposed to be productive. Added to it, mean depth of a reservoir may provide additional information in this respect. Characteristics of some reservoirs are indicated in Tables 11A and B.

Random sampling made on shore and volume development index of reservoirs of different sizes spread over the country together with their mean depth indicate high shore development ranging from 1.5 to as high as 13.5. Higher values recorded at FRL were 13.5 and 12.46 in case of Hirakud (71963 ha) and Govind Sagar (16867 ha). This is followed by Nagarjuna Sagar 28475 ha), Rihand (46538 ha), Krishnaraj Sagar (13200 ha), Tungabhadra Dam (37814 ha) and Mettur Reservoir (Stanley) (15346 ha) having their shore development index values as 7.89, 7.04, 7.25, 5.251 and 6.7. All these reservoirs having water spread area at FRL are above 10,000 ha. Contrary to such high value (shore development index value) in large reservoirs low value of 2.48 is also recorded in case of Pong Reservoir (24529 ha) Table 11 (A).

In case of smaller reservoirs (water spread area at FRL, 1000–10,000 ha) such values are also recorded. For example, Tilaiya (5921 ha), Konar (2590 ha), Santanur (2010 ha) and Bhavani Sagar (7876 ha) are indicating shore development index values as 9.1, 8.78, 6.2 and 4.0. Even in case of a very small reservoir, Sandynulla (258 ha) the shore development index recorded is 4.5. The rest however varied between 1.5

Table 11: Characteristics of Reservoirs
(A) Morpho-ecological Condition

Name of Reservoir with Area in ha at FRL	State	Year of Commission	Shore Development Index	Volume Development Index	Man Depth (m)	Temperature °C	Transperancy cm	pH	Dissolved Oxygen (ppm)	Alkalinity (ppm)	Specific Conductivity µ mhos	Nitrate (ppm)	Phosphate (ppm)	Silicate (ppm)	Gross Primary Production	Average Production t	Referece
10,000 ha and above																	
Hirakud (71,963)	Orissa	1957	13.5	0.52	11.3	–	39.6	8.2	9.6	39.6	151	0.05	0.04	7	–	350 (1985–1991)	Sugunan, V. (1995)
Gandhi Sagar (66,538)	Madhya Pradesh	1960	4.78	0.601	11.73	2.7–32.5	82–124	8.2–8.9	6.4–11.8	82.124	121	Tr–0.9	Tr–0.3	–	441–490	738 (1979–81)	Dubey, G.P. (MS)
Rihand (46,538)	Uttar Pradesh	1962	7.04	0.89	22.8	19.9–31.1	28–60	7.0–8.8	3.0–10.0	28–60	92.24	0.15–0.70	Tr–0.65	14	37.17–1022	167 (1971–1981)	Sugunan, V. (1995)
Tunga–bhadra (37,814)	Karna-taka	1953	5.251	0.737	9.8	23.1–29.5	30–100	7.9–8.5	4.8–11.5	30–100	240–359	0.2–0.5	Tr–3.3	9–16	194.4–2271.6	3085 (1980–1981)	Sugunan, V. (1995)
Nagarjuna Sagar (28,475)	Andhra Pradesh	1967	7.89	1.13	29.9	22.7–31.0	70–169	7.6–8.6	5.12–8.9	69.98–169.27	219.6–1114.7	0.2–3.2	0.001–0.003	19–40	90–1205	118.79 (1971–1980)	Final Report CICFRI (1983)
Pong (24,000)	Hima-chal Pradesh	1974/75	2.48	–	35.7	22.3–25.1	54–81.5	6.5–9.0	9.1–9.9	53.9–81.5	177.3–185.9	–	–	–	–	484.8 (1976–1990)	Sugunan, V. (1995)
Govind Sagar (16,867)	Hima-chal Pradesh	1963	12.86	1.01	55	9–31	20.7–59.9	7.3–8.4	1.76–13.44	50–104	119–291.6	Tr–0.15	Tr–0.12	0.8–9.6	135–1875	556.91 (1970–1993)	Sugunan, V. (1995)

Contd...

Table 11(A)–Contd...

Name of Reservoir with Area in ha at FRL	State	Year of Commission	Shore Development Index	Volume Development Index	Man Depth (m)	Temperature °C	Transperancy cm	pH	Dissolved Oxygen (ppm)	Alkalinity (ppm)	Specific Conductivity μ mhos	Nitrate (ppm)	Phosphate (ppm)	Silicate (ppm)	Gross Primary Production	Average Production t	Referece
Mettur/ Stanley (15.346)	Tamil Nadu	1939	6.7	1.4	18.3	24.2–32	140–278	7.5–8.8	4.5–12.6	140–278	170–350	Tr.	Tr–0.1	0–19.9	–	344 (1967–1977)	Sugunan, V. (1995)
Krishnaraj Sagar (13,000)	Karnataka	1932	7.25	0.78	30.17	20.1–29.8	36–140	7.2–8.4	6.0–8.0	36–140	–	Tr–0.011	Tr.–0.05	8–12	–	161.93 (1980–81)	Sugunan, V. (1995)
Below 10,000 ha																	
Ravi–sankar (9540)	Madhya Pradesh	1978	2.0	–	9.53	21.0–30.5	61.166	7.4–10.2	6.76–10.2	40–94	44–100	0.05	0.04–0.16	–	–	31.35 (1987–93)	Sugunan, V. (1993)
Haema–vathy (9162)	Karnataka	1981	1.6	0.754	–	22.5–29.75	32.5–179.5	7.2–8.2	4.5–8.2	23.43	–	–	0.02–0.05	0.05–0.08	–	130.0 (1984–86)	Sugunan, V. (1993)
Bhavani Sagar (7876)	Tamil Nadu	1953	4.01	1.0	11.9	23.7–30.2	18.5–135	7.8	7.48	40.92	269.96	Tr–0.34	0.01–0.06	0.14.3	–	193–485 197–84	Devaraj, S.P. *et al.* (1990)
Tilaiya (5921)	Bihar	1952	9.1	0.766	6.65	24.5–33.0	31.5	6.7–7.0	4.7–8.3	53.5	98.7–211.4	0.25	0.03	–	199–300	3.5 (1969–79)	Jhingran, V.G. (1975)
Getalsud	Bihar	1971	3.85	0.6839	32.6	16.0–28.5	14–176	7.3–8.2	6.4–9.6	–	–	Tr.–0.36	0.001–0.12	2.6–18.2	257.85–295.53	117.9 (1977–81)	Mar 1983, CICFRI

Contd...

Table 11(A)–Contd...

Name of Reservoir with Area in ha at FRL	State	Year of Commission	Shore Development Index	Volume Development Index	Man Depth (m)	Temperature °C	Transperancy cm	pH	Dissolved Oxygen (ppm)	Alkalinity (ppm)	Specific Conductivity µ mhos	Nitrate (ppm)	Phosphate (ppm)	Silicate (ppm)	Gross Primary Production	Average Production t	Referece
Konar (2950)	Bihar	1955	8.78	0.85	12.98	17.6–28.9	–	–	–	43.5–83	81.20–147.6	0.10–0.60	Tr–0.047	1.3–1.9	199–300	3.71 (1968–71)	J.V.G. and A.V.N. 69
Sathanur (2010)	Tamil Nadu	1957	6.2	1.0	10.1	25.0–30.4	–	8.3	8.5	145–616	320–800	4.5–5.6	–	0.1–3.2	–	125.688 (1966–84)	D.M.G. *et al.*, 90
Ramgarh (1260)	Rajasthan	1963	1.81	0.61	4.6	18.0–32.0	80–108	8.3–8.6	6.3–9.2	129–198	307–791	0.22–2.13	Tr.–0.4	5.6–18.46	–	70.734 (1969–85)	Dubey (MS)
Krishnagiri (1248)	Tamil Nadu	–	1.5	0.94	5.2	23.5–31.4	11.5–70	8.0	6.8	146–373	425–575	–	–	0.34–4.4	–	47	Sugunan, V. (1993)
Amaravathy (850)	Tamil Nadu	1959	2.3	1.2	13.7	24.0–28.8	–	6.7–9.1	8.1	7–84	38–63	1.7–2.8	0–0.01	1.4–38.5	–	71.520 (1961–84)	Devraj, S.P. *et al.* (1990)
Tirumurthy (460)	Tamil Nadu	–	–	1.0	11.0	–	–	7.7	7.6	16–124	40–200	–	0–2.0	5–16	–	13.0	Sreenivasana, A. (1995)
Sandynulla (258)	Tamil Nadu	–	4.6	1.2	10.3	16.4–22.6	78.5	7.95	8.3	26–62	99–280	1.5–2.1	Tr.	0–4.7	–	5.0	Sreenivasana, A. (1978

Contd...

Table 11(B): Photosynthetic Energy Fixation and Energy Conservation in Reservoir Biotops

Name of Reservoir with Area in ha at FRL	Total Radiont Energy g.Cal/m²/yrx10⁵	Photosynthetic Production Energy Cal/m²/yrx10⁵	Efficiency of Conservation of Radient Energy to Chemical Energy (%)	Fish Production		Conversion of Energy		Referemce
				g/m²/yr.	As Energy Cal/m²/yr	Fish/ Photosynthetic	Fish/ Light	
10,000 ha and above								
Rihand	6,877	1.387	0.202	0.44	528	0.038	0.000076	Natarajan, A.V. and V. Pathak (1983)
Nagarjuna Sagar	7,483	2.175	0.290	0.79	948	0.043	0.00013	–do–
Govind Sagar	6,278	4.269	0.682	5.36	6432	0.151	0.0010	–do–
Mettur/Stanley	7,775	8.147	1.047	6.29	7550	0.0927	0.000971	Jhingran, V.H. (1975)
Below 10,000 ha								
Bhavani Sagar	7,775	3.205	0.412	5.94	7128	0.220	0.00091	Natarajan, A.V. and
	7,775	8.162	10.50	2.78	3336	0.0649	0.00194	V. Pathak (1983)
	7,775	8.162	1.050	3.32	3792	0.0775	0.000613	
Sautanur	7,738	8.100	1.047	7.70	9241	0.1141	0.00194	Jhingram, V.G. (1975)
Ramgarh	6,680	3,007	0.49	0.821	9853	0.33	0.00146	Sugunan, V. (1995)
Krishnagiri	7,720	5.610	0.727	2.05	2460	0.1449	0.000319	Jhingran, V. G. (1975)
Amaravathy	7,847	11.320	1.442	16.00	19230	0.1698	0.002451	–do–
Trimurthy	7,845	14.15	1.803	3,456	4147	0.02931	0.000529	Sugunan, V. (1995)
Sandynulla	8,578	8.121	0.947	4.54	5454	0.0671	0.000636	Jhingran, V. G. (1975)

and 2.96. In short, it can be said that moderate to high shore development index are common in case of most of the medium size reservoirs [Table 11(A)].

The convex lake basin wall (volume development index <l) are also significant and is evidenced from their values 0.53–0.89 in case of reservoirs 20,000 ha and above. While in comparatively smaller reservoirs (10,000–20,000 ha) their ranges are 0.6–0.94. In another smaller reservoir (773 ha) the volume development index recorded is 0,68. Concave basin wall (volume development index> 1) is also not uncommon, but this is recorded mostly in case of reservoirs less than 1000 ha at FRL. Their values ranged between 1.0–2.0 [Table 11(A)].

The study made in two multipurpose reservoirs, Tungabhadra and Nagarjuna Sagar whose shore development index were 5.251 and 7.89 respectively have provided information on the areas of shallow littoral zones that exist at various levels during different months. Table 12 indicates depth of water under 3 and 1.5 m. for Tungabhadra which is biologically productive and remains constant except at 484.8–481.8 m when there is a sharp fall in the available shallow areas. Areas under 3.0 m of water vary from 15.6–18.3 per cent of the total reservoir area between 439.9 and 481.8 m levels and a maximum from 29.6 to 30.6 per cent between 487.8 and 486.3 m levels (Table 12).

Table 12: Fluctuations in the Water Level Affecting Changes in the Hydrography of Tungabhadra Reservoir

Contour Level (Meter Abve m.s.l.)	*Period 1965*	*Total Water Area (sq.km.)*	*Per cent Exposed*	*Area (sq. km.) Under*	
				1.51 m Depth	*3.03 Depth*
493.9	January	356.90	-	32.375	65.527
492.4	February	324.27	9.0	32.893	68.117
490.9	March	256.15	19.0	35.250	68.117
489.4	March	256.15	29.0	37.379	68.637
487.8	April	220.67	38.0	33.281	65.268
486.3	April	187.52	48.0	31.935	57.498
484.8	April	155.66	57.0	25.511	46.361
483.3	May	130.02	64.0	20.953	38.021
481.8	June	109.00	72.0	–	–

Source: Ray, P. and R.K. Banerjee (1969).

In case of Nagarjuna sagar based on Final Report 1971-81 on Nagarjuna Sagar, Inland Fisheries Research information Series 3, March 1983. Barrackpore the extent of shallow areas (under 5 ft and 10 ft depth) in relation to reservoir level between 480' and 590' are given in Table 13 as the minimum level obtained till now at FRL are in between this range. From this table it can be seen that the shallow areas under 5 ft never exceeded 3.9 per cent of the total area and those under 10 ft never exceeded 7.5 per cent of any level of reservoir (Table 13). However, these shallower areas almost form a constant parcentage except at highest levels of 585 ft and 590 ft. Thus productive area almost remained constant throughout the year. The litteral development under 30 ft. area was between 13 and 18 per cent of the total area between these levels. The above results indicate that a good percentage of the reservoirs remain shallow round the year and can be biologically productive. But, during late or non-monsoon months the inflow supply from the river may not cope with the outflow of the reservoir, as a result the water level of the main reservoir shows a gradual fall and the productive shallow zones gradually get exposed. Such phenomenon occurred in case of Tungabhadra Reservoir (Ray and Banerjee, 1969).

Table 13: Extent of Shallow Areas of Nagarjuna Sagar in Relation to Reservoir Levels

Reservoir Level (f)	*Area Under 5 ft. (ha)*	*Area Under 10 ft. (ha)*	*% of Area in Total Area*	
			5 ft	*10 ft.*
590	261.6	523.3	0.9	1.8
585	261.7	1,167.6	0.9	4.1
580	905.9	1,811.9	3.2	6.5
575	906.0	1,600.2	3.3	5.9
570	694.2	1,388. 4	2.7	5.3
565	694.2	1,447.9	2.7	5.7
560	753.7	1,507.4	3.0	6.1
555	753.7	1,501.7	3.1	6.3
550	748.0	1,496.0	3.2	6.4
545	748.0	1,301.0	3.3	5.8
540	553.0	1,1C5.6	2.5	5.1
535	552.8	1,087	2.6	5.1
530	534.2	1,068.4	2.6	5.2
525	534.2	1,029.6	2.7	5.1
520	495.4	990.9	2.5	5.1
515	495.5	1,071.5	2.6	5,6
510	576.0	1,152.0	3,1	6.2
505	576.0	1,231.7	3.1	6.2
500	655.7	1,311.4	3.8	7.5
495	655.7	1,078.8	3.9	6.4
490	423.1	846.2	2.6	5,2
485	423.1	1	2,7	–

Final Report 1971-81 CIFRI series 3, March 1983.

Dubey (1969) in his study on "Development of reservoir fisheries in MP, "clearly demarketed the productive areas in respective of full reservoir levels for seven reservoirs and has termed them as developed reservoirs (Table 14). The percentage productive areas are also shown in the table.

Table 14: Showing Productive Areas in Respect of Total Reservoir Area at FRL of M.P.

Name of Reservoir	*Location*	*Area (ha) at FRL*	*Productive Area*	*Percent-age Area*	*Average Production/ ha (kg.)*	
					FRL	*Productive Area*
Harsi (D)	Gwalior	2560	965	37.6	20	52.8
Pargana (D)	Morena	1441	480	33.3	18.6	55.9
Madan Sagar Jatara (D)	Tikamgarh	444	200	45	54.2	120.4
Gwalsagar (D)	–Do–	280	212	95.9	153.7	203
Yeshwant Sagar (D)	Indore	1025	460	44.8	22.9	51.1
Maniari (ND)	Bilaspur	2400	800	33.3	15.1	45.3
Gandhi Sagar (ND)	Mandsaur	68000	34000	50	3.3	6.6

Nos. 6 and 7 are newly developed and the rest are well developed. Production figure is based on average of 1963-64 to 1966-67.

Chapter 10

Transition from River to Reservoir: Their Impact on General Ecology and Biomass

Dams bring about considerable morpho-ecological changes in the ecosystem. These are:

1. The flow of river is reduced considerably.
2. With the inundation of the catchment area the water initially becomes turbid and heavy silt precipitation occurs from erosion of the catchment area.
3. Initially a change gradually occurs in the physico-chemical character of the water with the enrichment of the water soluble nutrients available from the washings of the eroded silt of the catchment area.
4. The flora and faunal composition with species (quantitatively and qualitatively) of the main river is gradually changed which is more suited to the reservoir water after it attains a sort of lentic conditions. The benthos adopted in the deeper water is different from shallow zone or from lotic condition of river water. Benthic insectivores are adversely affected. Increased sedimentation reduces insect diversity, density and species richness (Chutter, 1969; Gammon, 1970; Nuttall and Bielby, 1973; Brusven and Prather, 1974).
5. The inundation of natural spawning ground of commercially important species affect their population. Fish may be more affected by alter nation in breeding ground and spawning condition than by alterations in their food (Balon, 1975). Siltation generally degrade spawning grounds, cause behavioural changes in spawning fish, increase egg mortality, decreases

larval growth and affects development rate and survival of fish (Langlois, 1941; Cordone and Kelley, 1961; Smith, 1971; Nancy *et al.*, 1979).

6. The regulated discharge of water (due to construction of barrage, Dam etc.) lowers the magnitude of flow downstream for most part of the year which encourages recurring annual silt precipitation, lowering of depth, increase flood (Table 5) due to overflowing or breaching of banks, and in the estuarine zone (Bhagirathi-Hooghly) encourages the migration of brackish-water fishes further up along with the increase in salinity during tidal influx. Migratory species like hilsa and the freshwater prawn-Macrobrachium spp. are greatly affected and deprived of their breeding runs in the river due to faulty design of the barrage.

Chapter 11
Physical and Chemical Status

Data presented under this heading is the average value obtained from various points during different periods for individual reservoirs large, medium and small. Further, effort is also made to highlight the characteristics of large (above 10,000 ha) and medium size (below 10,000 ha) reservoirs that indicated fairly productive (Table 11(A). Data on bottom waters are not presented here except in a few instances. Some interesting changes in the Physico-chemical conditions observed in case of Gitalsud and Rihand Reservoir of North India and for Tungabhadra Dam, Nagarjuna Sagar and Bhavani Sagar Reservoir of tropical zones are also indicated in Figures 7(a-f), 8 (a-c) and 9, 10 (a-c) and 11 (a, b). Heat dynamics in different reservoirs are indicated in Figure 12.

Temperature

The non-homogeneous temperature distribution with depth resulting from solar radiation, water surface heat temperature, adsorption, avection and heat by inflow, convective currents and other causes results density stratification in reservoirs. The temperature gradient being directly related to density gradient may become the governing parameter in the reservoir dynamics. This results high temperature at the surface layer which decrease with depth. The temperature of the epilymnion is about the same from top to bottom, varying a few degrees when surface water is warming rapidly in the absence of wind. Below the epilymnion is a layer of water (Themocline and metalimnion) where the temperature decrease rapidly proceeding downwards @ 1°C or more and below thermocline, the temperature appears to be almost uniform. Such thermal stratification results in a reduced heat budget for the reservoir, since heat is concentrated at the surface causing more evaporation and also heat loss. Due to separation of cold layer from the photosynthetic zone, algal and atmospheric

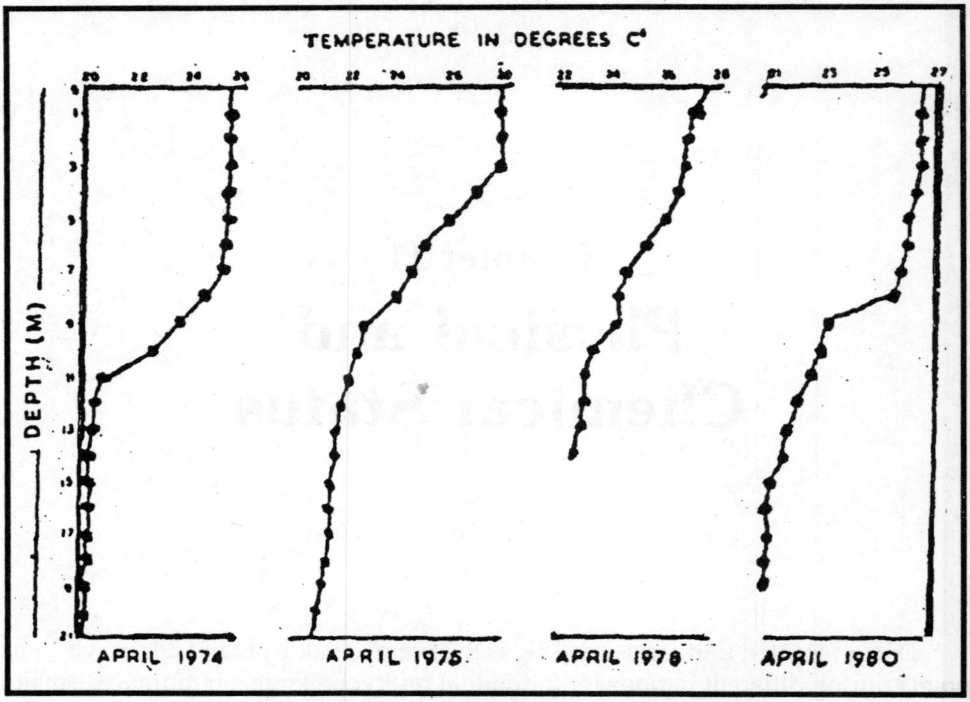

Figure 7a: Thermal Stratification in Gitalsud Reservoir
(*Source*: Final Report 1971–81, IFRI, Series 4, March, 1984)

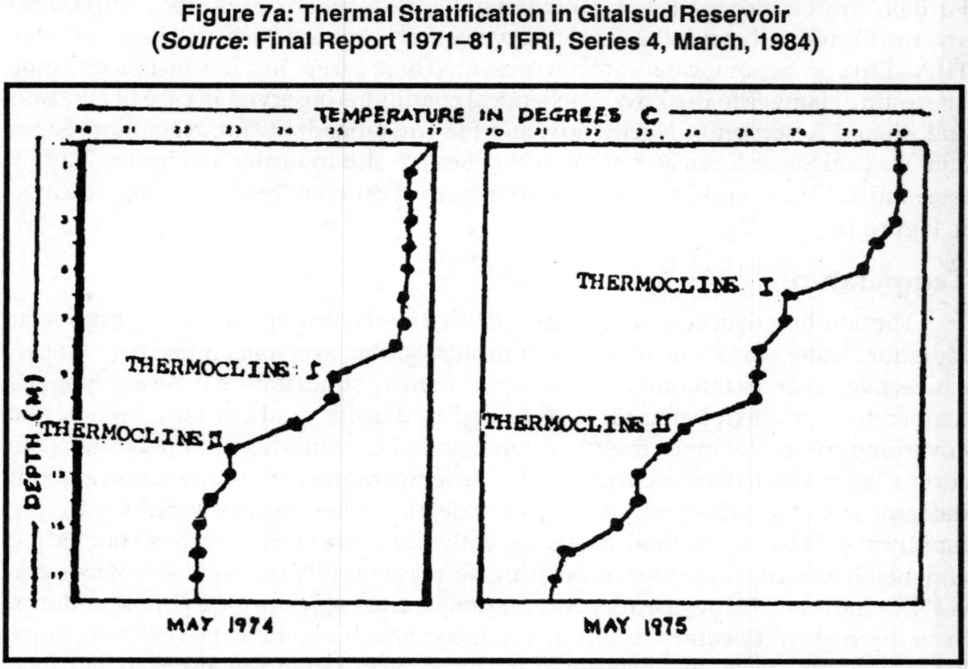

Figure 7b: Two Thermoclines in Depth Profiles of Gitalsud Reservoir
(*Source*: Final Report 1971–81, IFRI, Series 4, March, 1984)

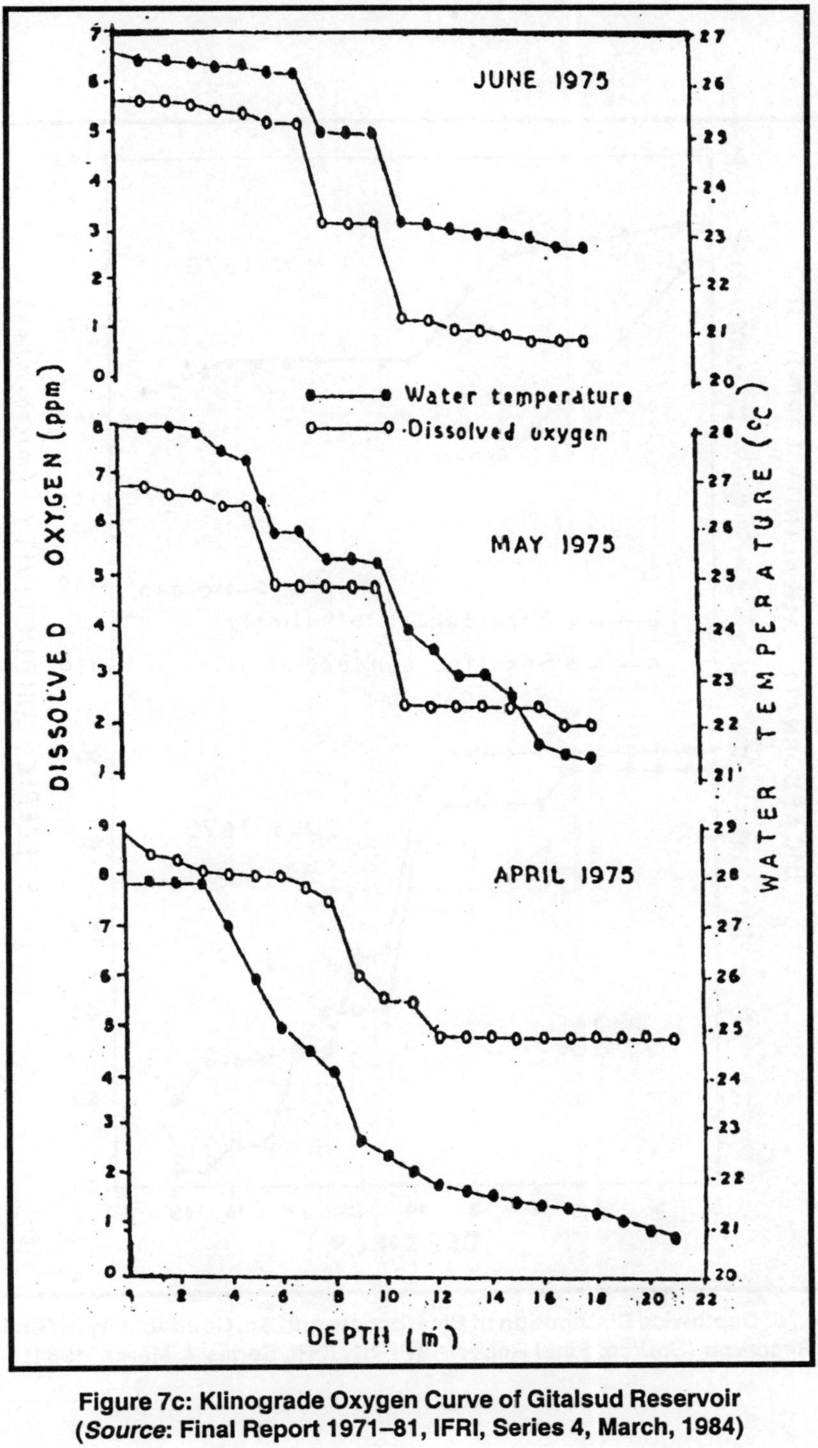

Figure 7c: Klinograde Oxygen Curve of Gitalsud Reservoir (*Source*: Final Report 1971–81, IFRI, Series 4, March, 1984)

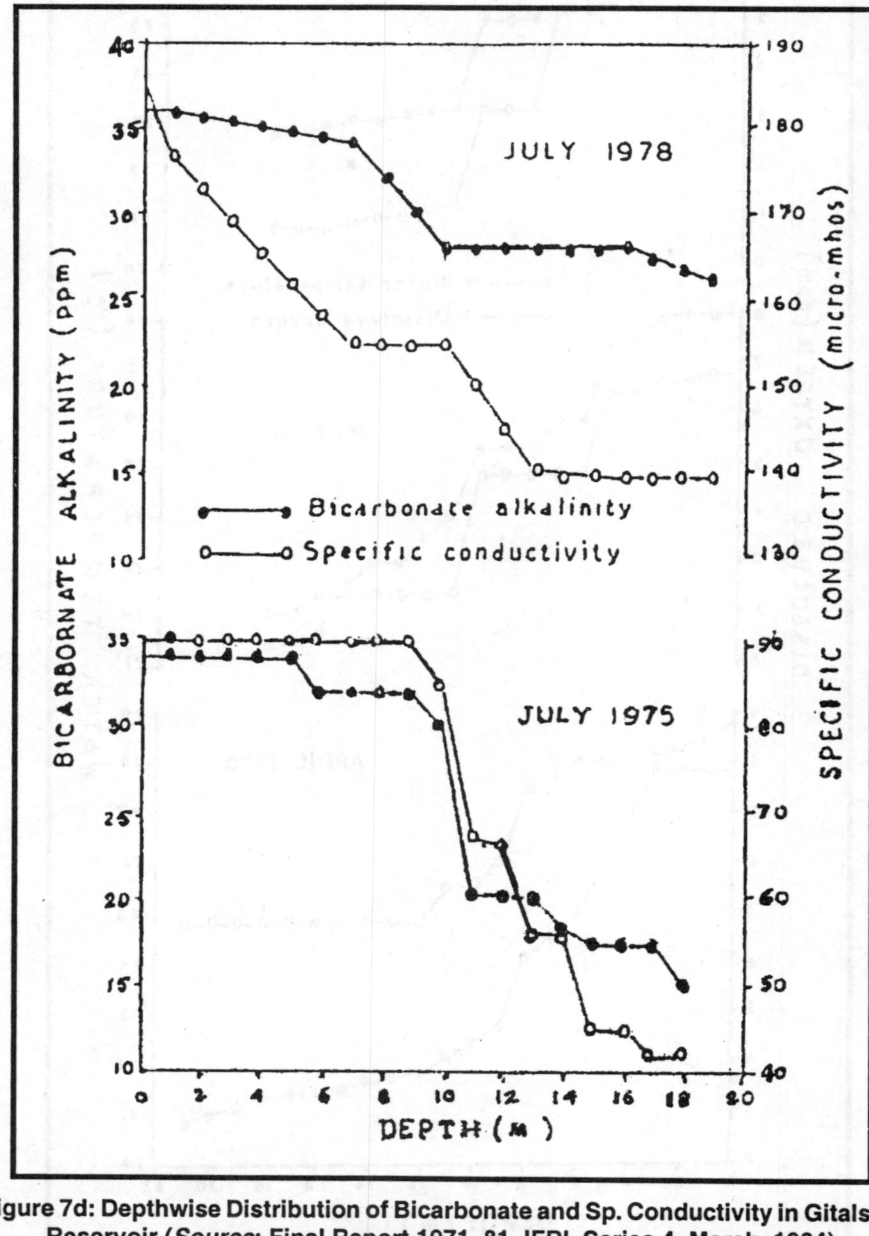

Figure 7d: Depthwise Distribution of Bicarbonate and Sp. Conductivity in Gitalsud Reservoir (*Source*: Final Report 1971–81, IFRI, Series 4, March, 1984)

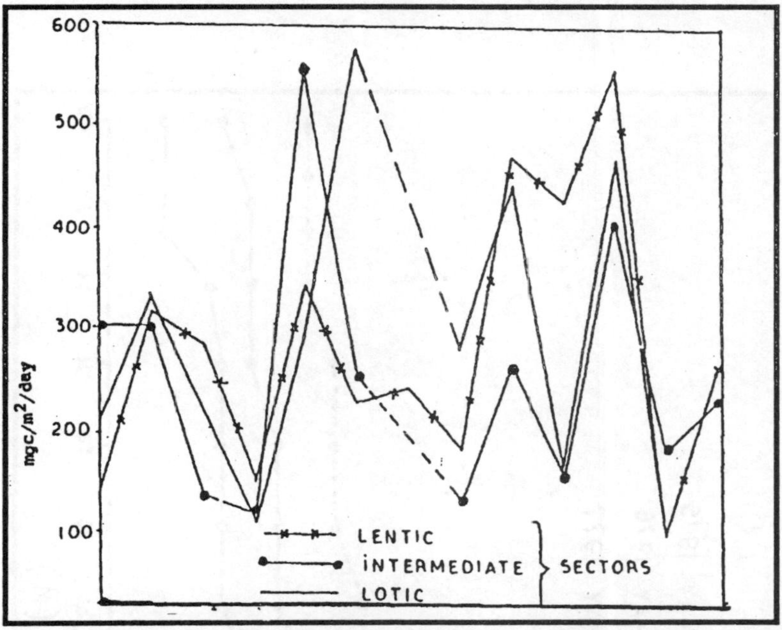

Figure 7e: Primary Production in Different Season of Different Years of Gitalsud Reservoir (*Source*: Final Report 1971–81, IFRI, Series 4, March, 1984)

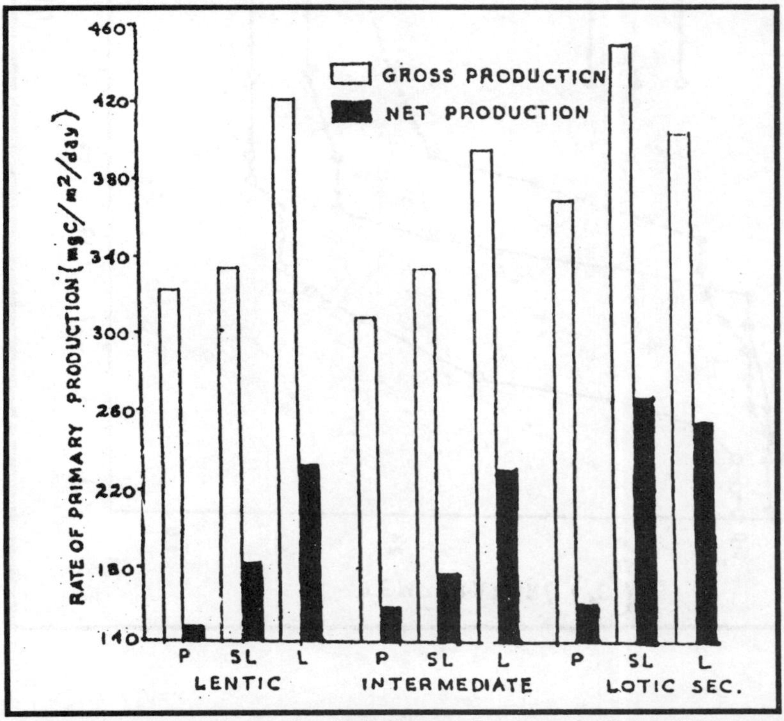

Figure 7f: Primary Production in Different Season of Gitalsud Reservoir (Av. value from October 1977–December 1980) (*Source*: Final Report 1971–81, IFRI, Series 4, March, 1984)

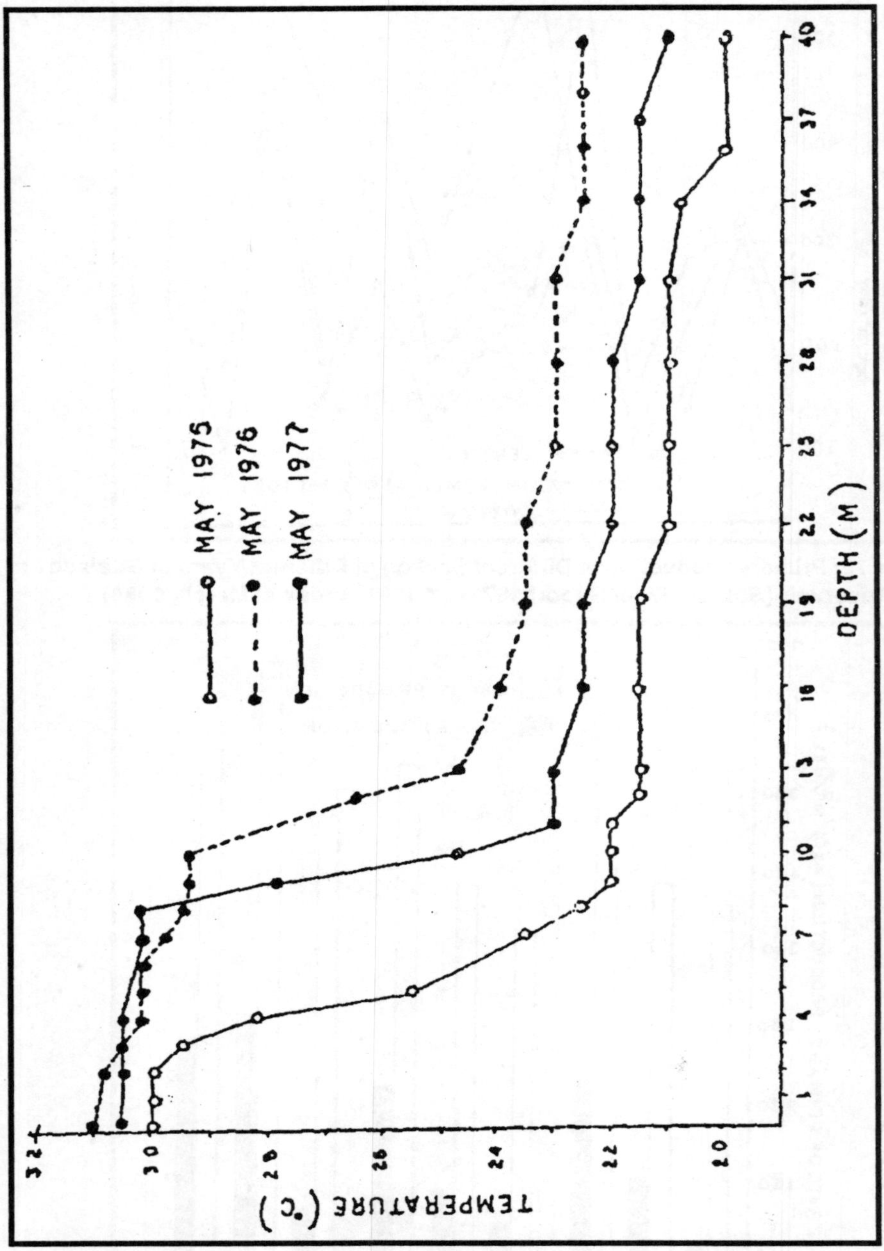

Figure 8a: Thermal Stratification of Rihnad Reservoir During Different Periods
(*Source:* Desai, V.R. and R.K. Singh, 1979)

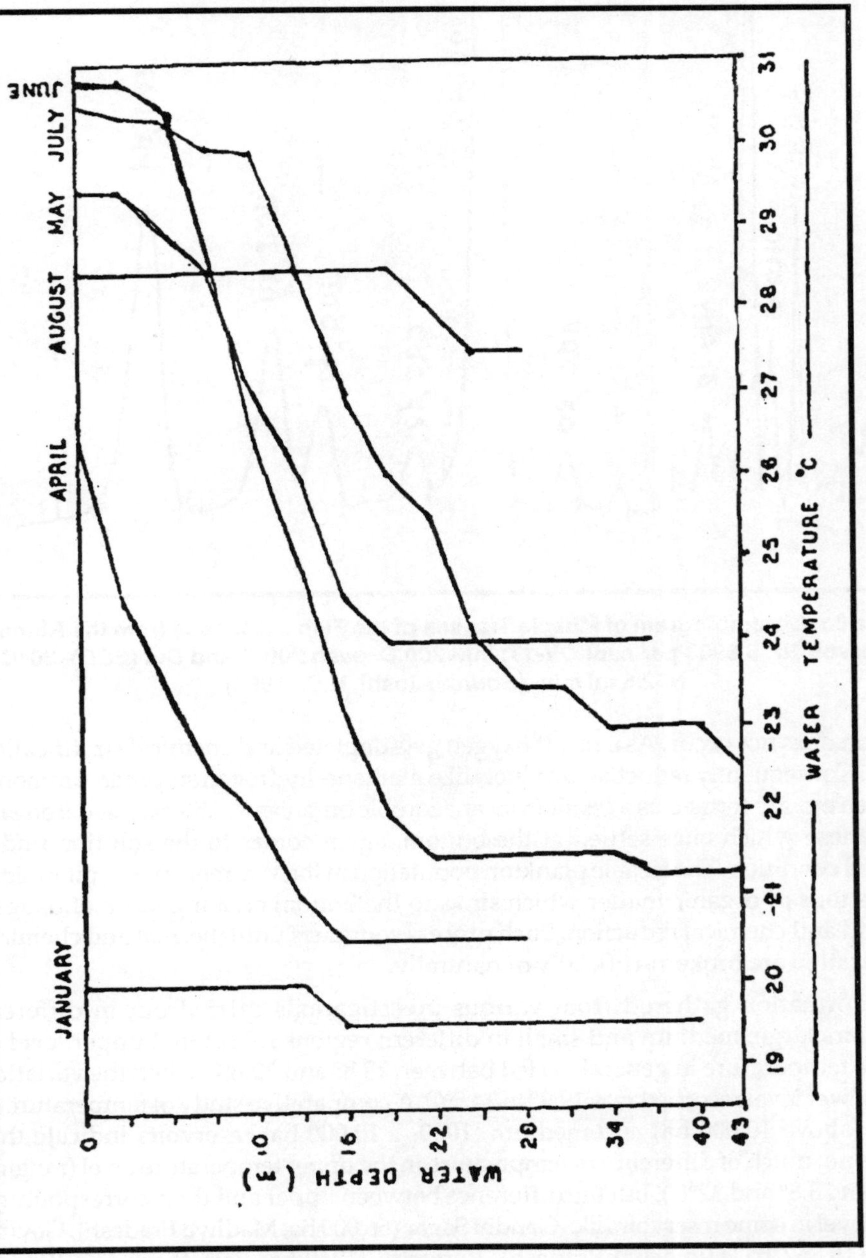

Figure 8b: Thermal Stratification of Rihnad Reservoir During Different Seasons
(*Source*: Final Report 1971–81, IFRI, Series 2, March, 1981)

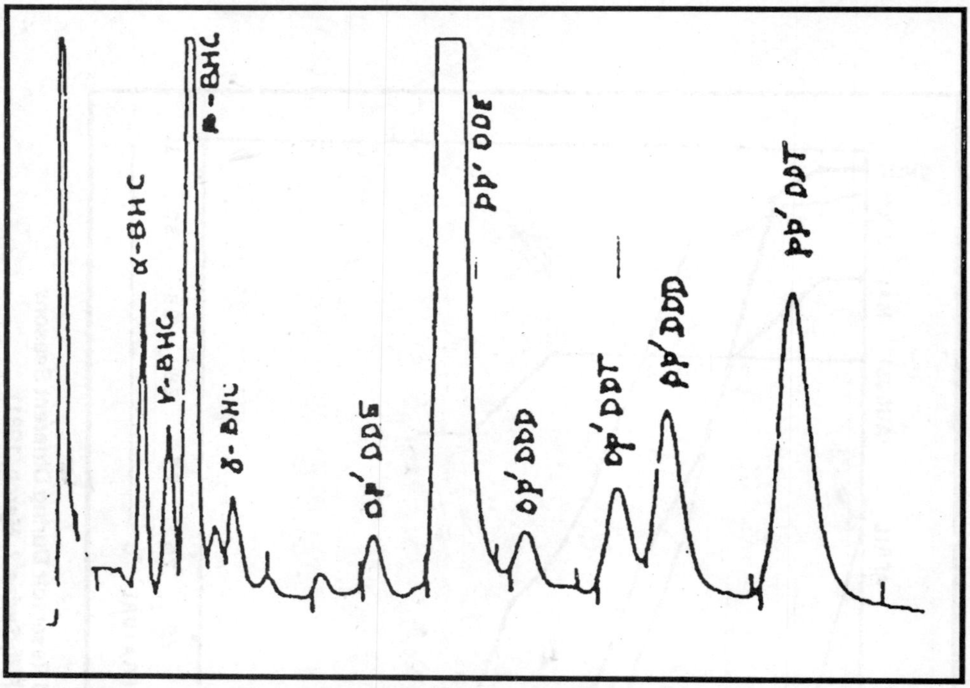

Figure 8c: Chromotogram of Muscle Tissues of the Fish *Catla catla* from the Rihnad Reservoir HP 5 8903 per cent OV-17; inj.- 200 C. oven 200 C and Det (ECD)- 30°C. N62.5 ml min. (*Source*: Joshi, H.C., 1990)

aeration does not occur. As a result oxygen gets depleted and chemical stratification sets in. Consequently reductive products like methane, hydrogen sulphide, ammonia nitrogen etc. are formed as a result from anaeorobic organisms. The oxidised iron and manganese which once settled at the bottom, again comes to the solution under reduced condition. The sizable plankton population in the Warmer layer (epilimnion) creates tons of organic matter which sinks to the bottom creating more of oxygen demand and chemical reduction. Such process continues until thermal and chemical stratification are broken artificially or naturally.

Information gathered from various investigations carried out in different reservoirs large, medium and small in different regions reveal that upper level of surface temperature in general varied between 28.8° and 32.6°C when the variation in the lower level recorded was 19.9° to 24.5°C. A comparative study of temperature of larger (above 10,000 ha) and medium (1000- ∠10,000 ha) reservoirs indicate that there is not much of difference of temperature in the upper temperature level (ranging between 28.8° and 32°C), but the difference between upper and their corresponding lower level in some reservoirs like Gandhi Sagar (66,000 ha, Madhya Pradesh), Govind Sagar (16,867 ha, Himachal Pradesh), Ramgarh (1260 ha, Rajasthan) and Rihand (46,538 ha. Uttar Pradesh) are significantly high (9°- 32.5°C). The range of lower level in general was showing variation between 19.9° and 24.5°C except in case of Gandhi

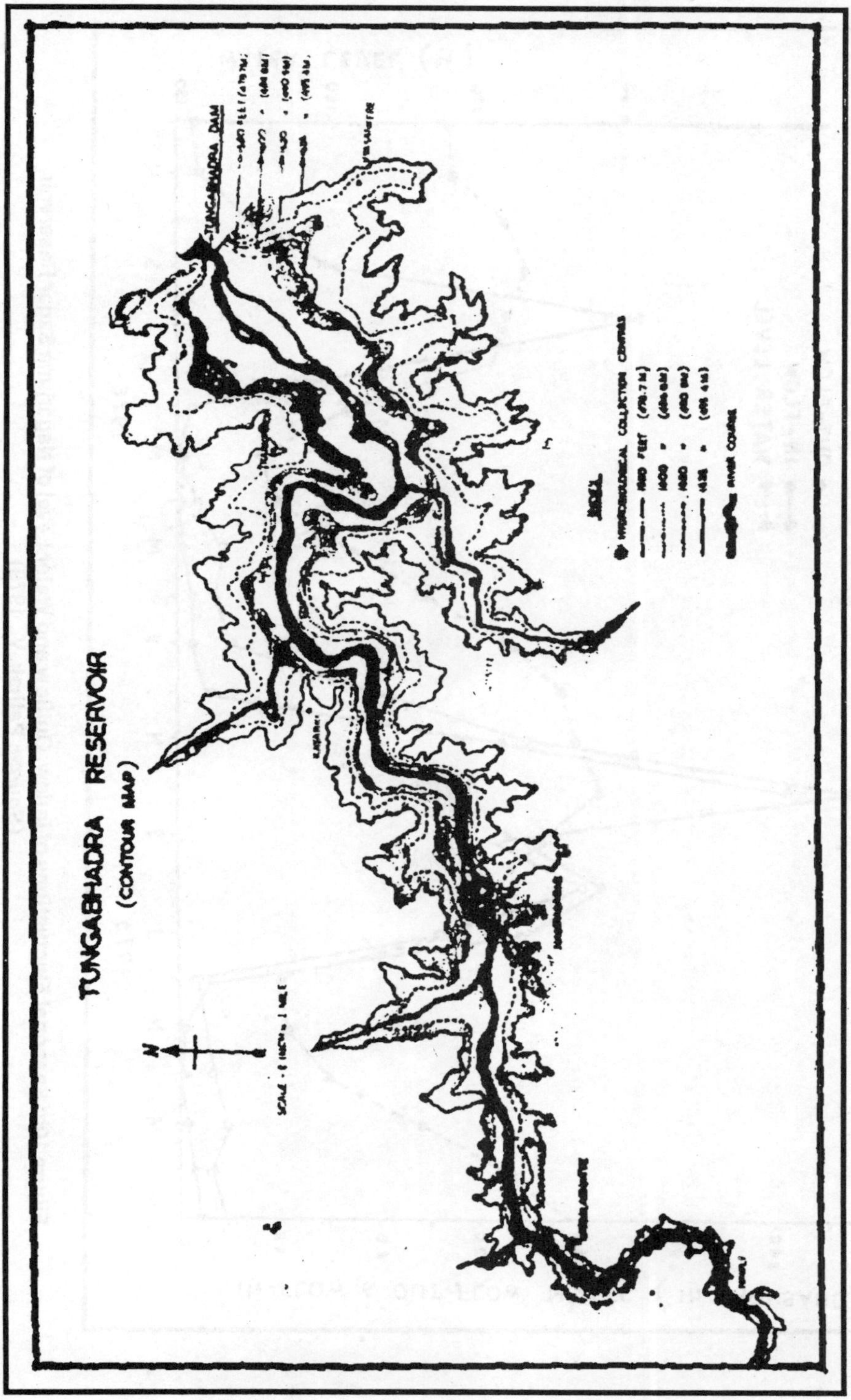

Figure 9: Showing Vast Exposed Areas (Littoral Zone) of Tungabhadra Reservoir (Karnataka) During Different Water Levels

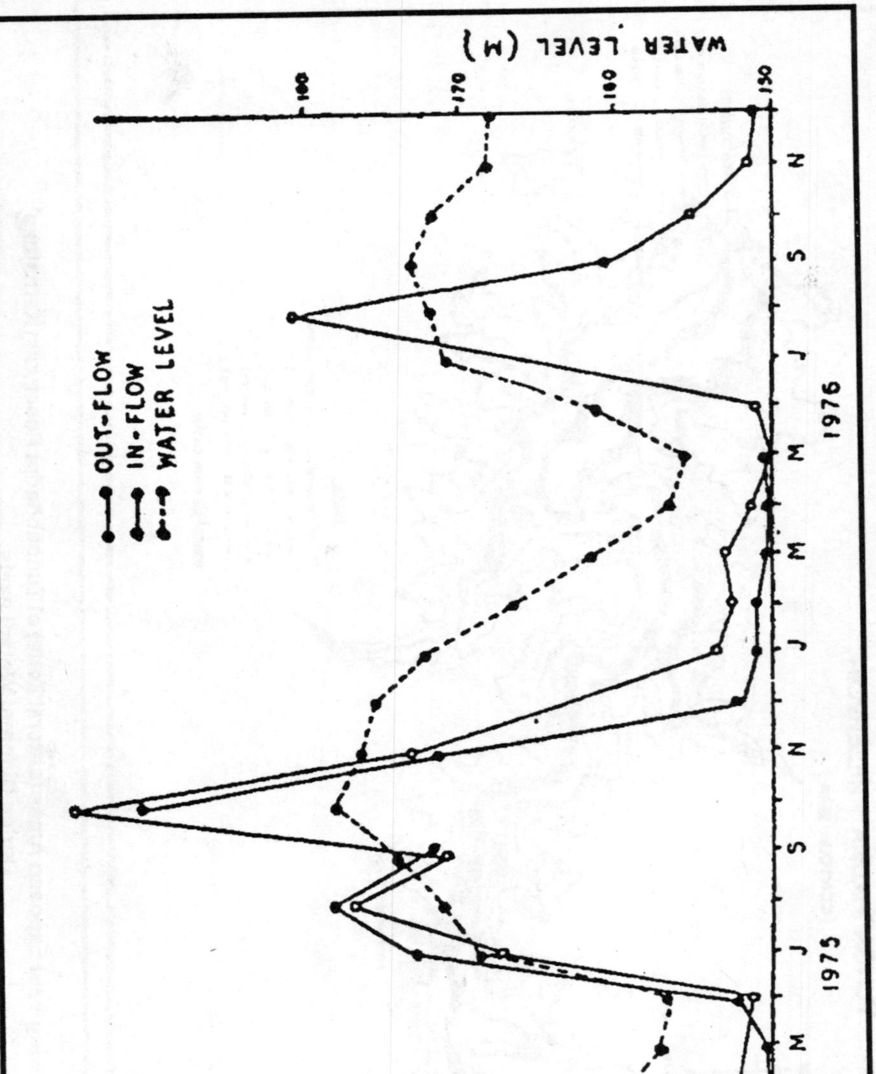

Figure 10a: Seasonal Fluctuations of Inflow, Outflow and Water Level of Nagarjuna Sagar Reservoir
(*Source*: Pathak, V., 1979)

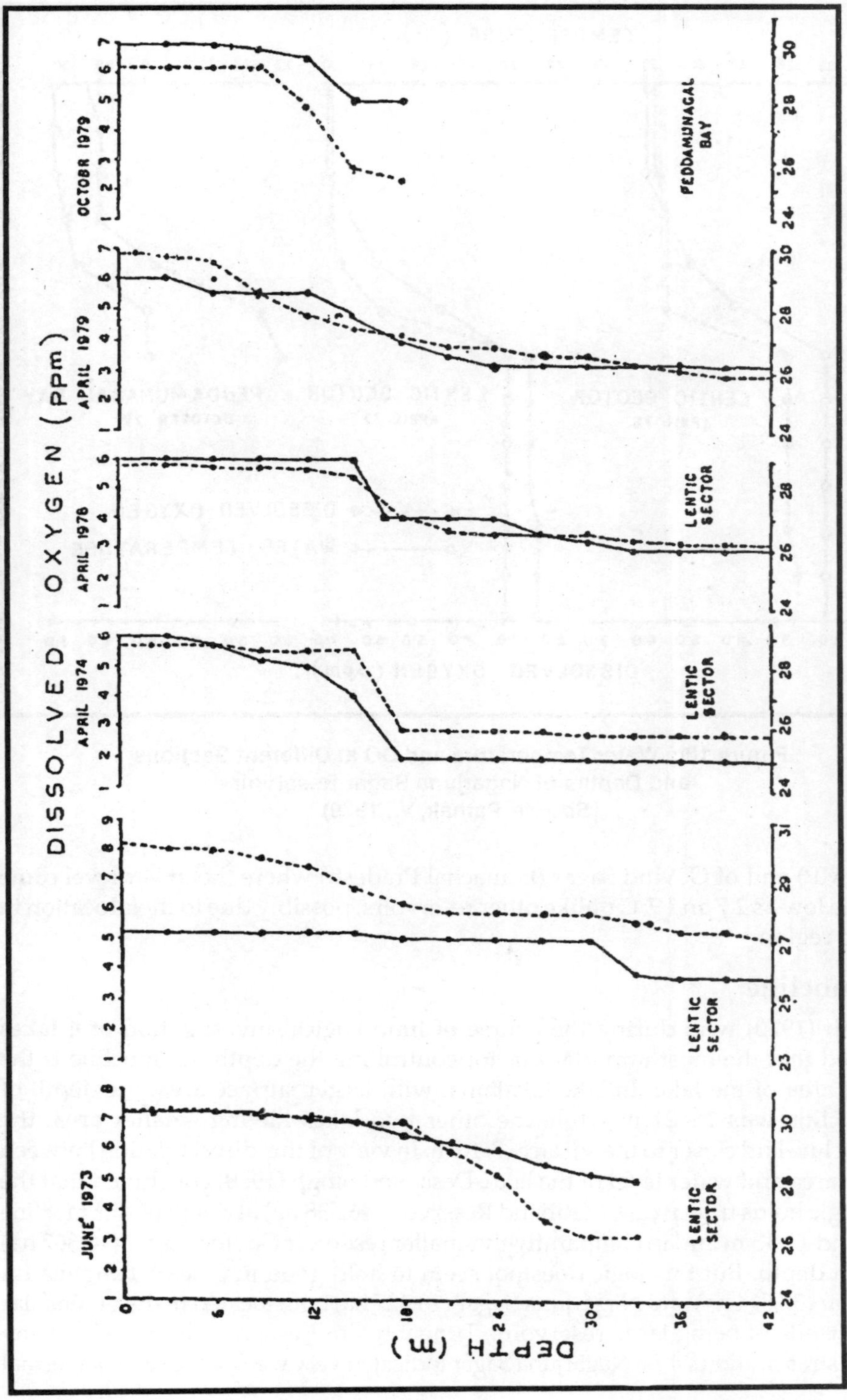

Figure 10b: Water Temperature and DO at Different Sections and Depths of Nagarjuna Sagar Reservoir (*Source:* Final Report, 1971–81, Series 3, March, 1983)

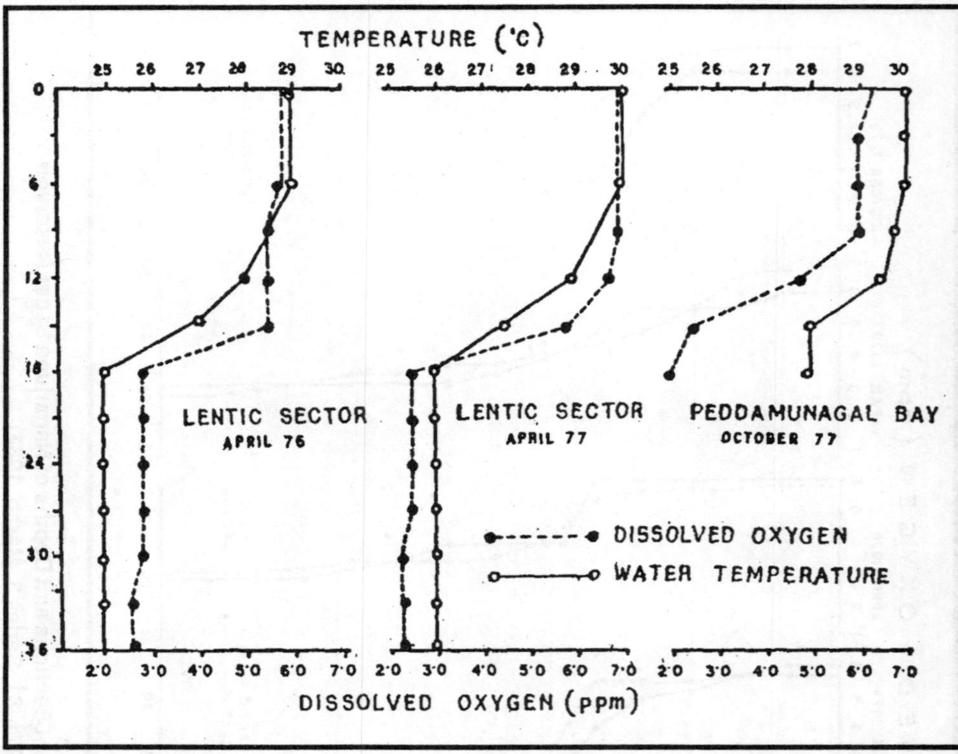

Figure 10c: Water Temperature and DO at Different Sections and Depths of Nagarjuna Sagar Reservoir (*Source*: Pathak, V., 1979)

Sagar (MP) and of Govind Sagar (Himachal Pradesh) where the lower level came down as low as 2.7 and 9°C unlike other reservoirs, possibly due to their location in the cold region.

Thermocline

Fish (1970) who during the course of limnological investigation of 4 lakes observed that the most important factor controlling the depth thermocline is the surface area of the lake. In lake Gkataina, with larger surface area, the depth of thermocline was 16–21 m while the other two lakes having smaller area, the thermocline laid closer to the surface–7-10 m. In view of this direct relation between surface area and water level of the lake, Desai and Singh (1979) concluded that the same logic holds true in case of Rihand Reservoir (46,538 ha) at depth of thermocline 5-7 m and 11-13 m and in comparatively smaller reservoir Govind Sagar (16,867 ha) at 2-6 m depth. But this logic does not seem to hold good in case of Tungbhadra Reservoir (37,814 ha) and Nagarjuna Sagar (28,427 ha) both located in the peninsular India, inspite of being large reservoirs, Tungabhadra Reservoir did not show any thermal stratification while Nagarjuna Sagar indicated very week indication of thermal

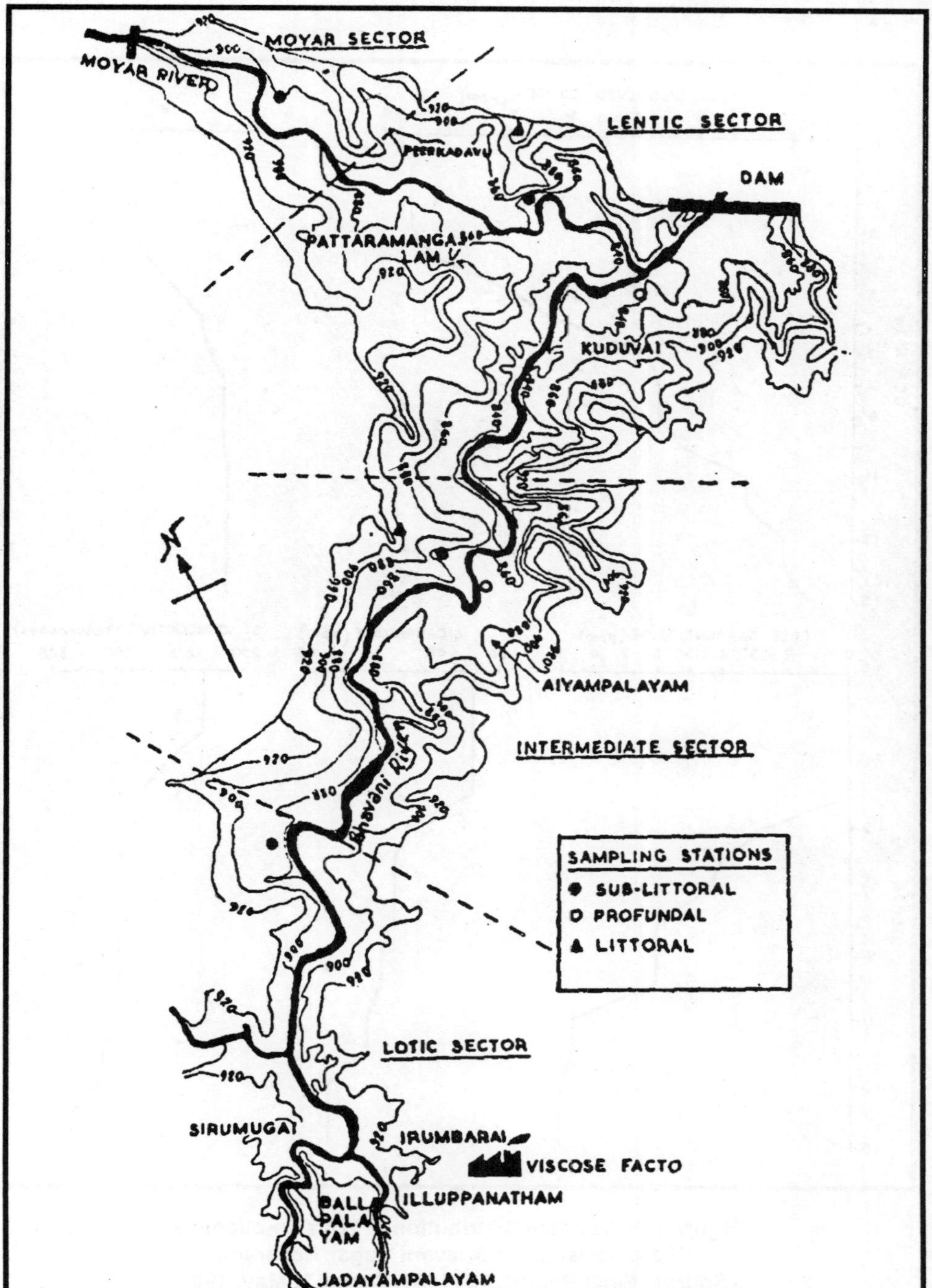

Figure 11a: Showing Vast Exposed Areas (Littoral Zone) of Bhavanisagar Reservoir (Tamil Nadu) During Different Water Levels

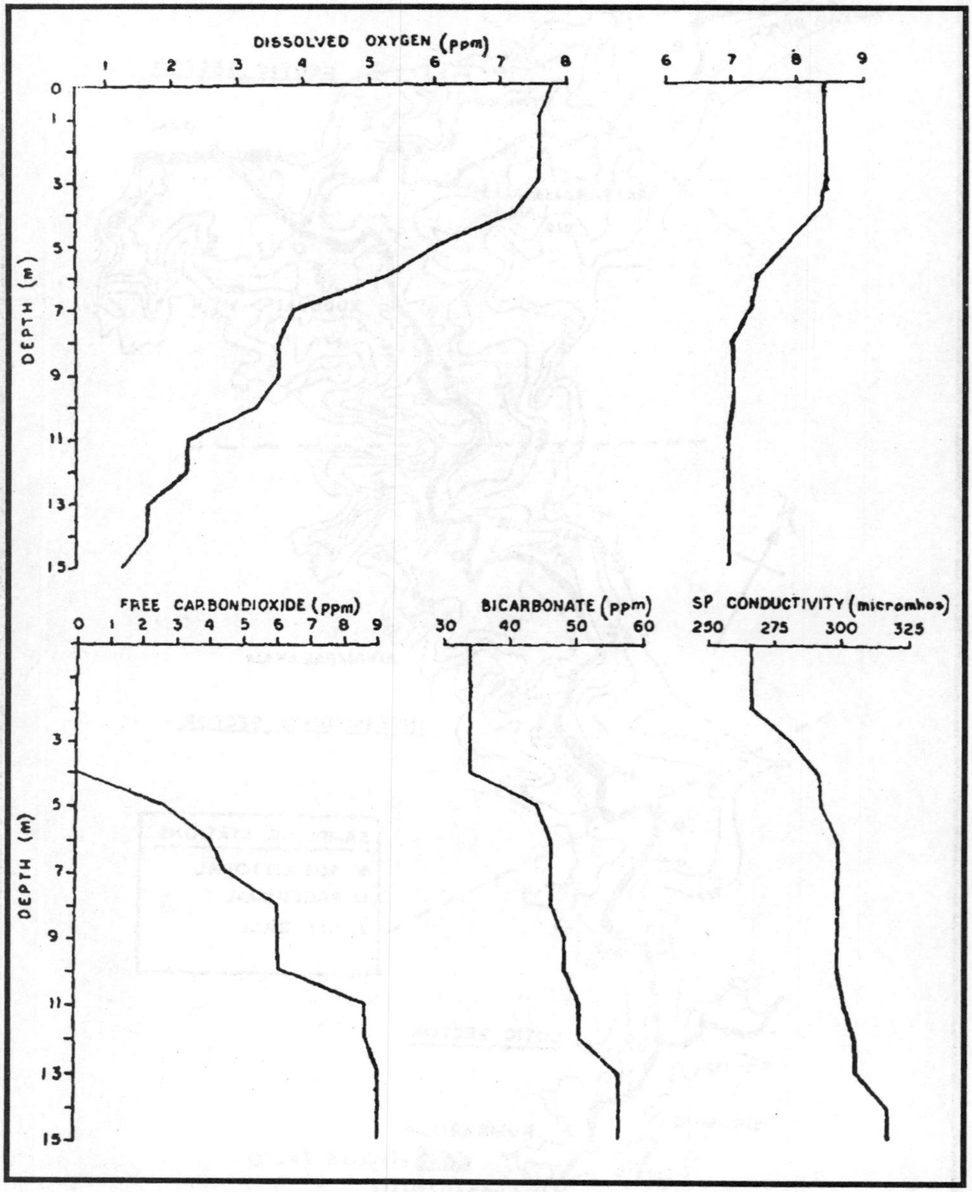

Figure 11b: Veritical Distribution of Physico-chemical Characteristics of Bhavani Sagar Reservoir (*Source*: Final Report, 1971–81, Series 1, May, 1981)

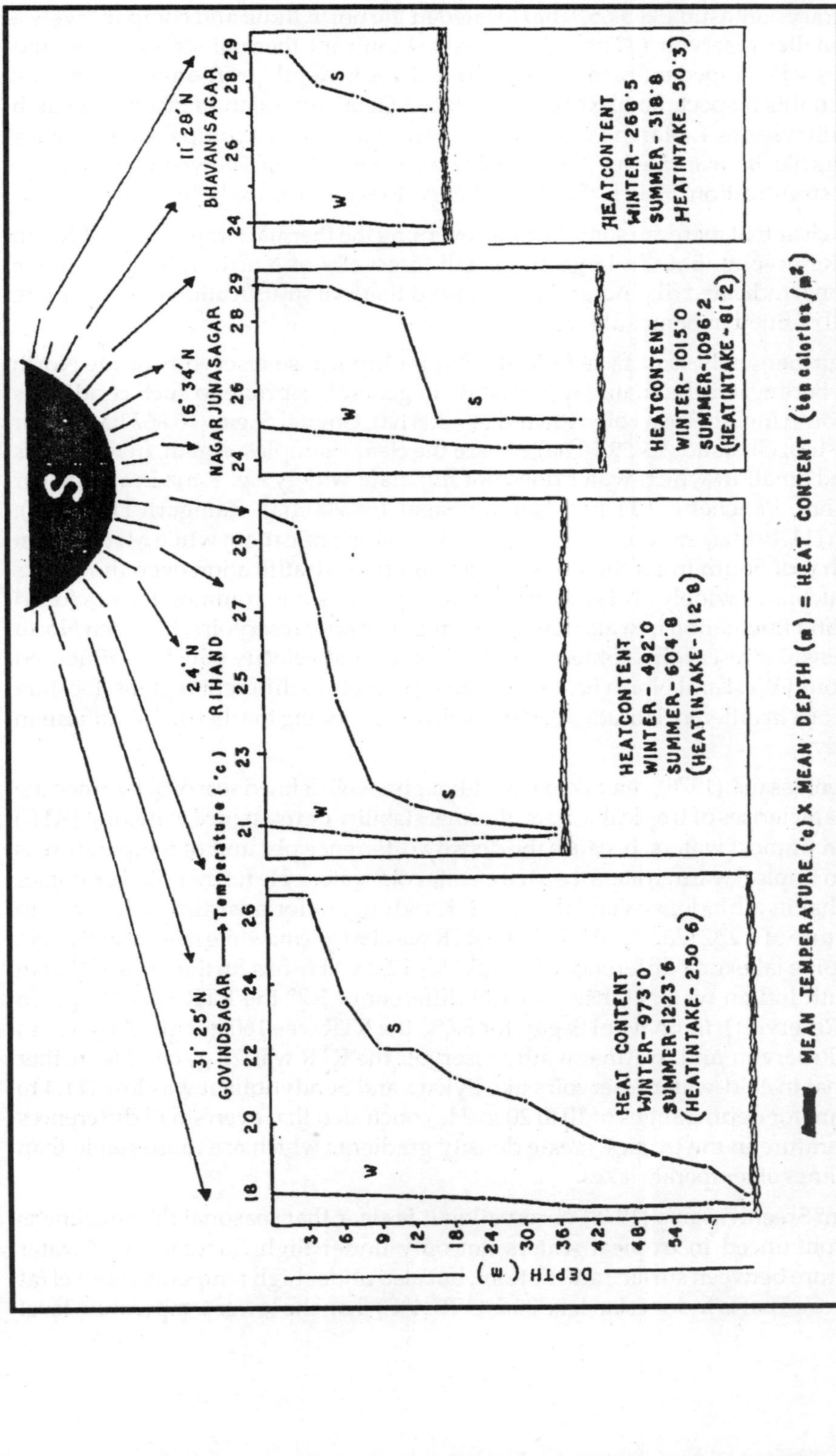

Figure 12: Heat Dynamics in Different Reservoir (ton calories/m²)
(*Source:* Natarajan, A.V. and Pathak, V., 1983)

stratification during summer (difference of 3-4 C was noticed during summer). On the contrary, Gitalsud (3,454,520 ha) located in the north India and comparatively a much smaller reservoir (3459 ha) indicated significant thermal stratification and sometimes developed more than one thermoclines in depth profile and is therefore unique in this respect. Final Report 1971-81 on Getalsud. Inland Fishery Research Information series 4. March 1984) Again contrary to this, in a comparatively larger reservoirs like Bhavani Sagar (7,876 ha) located in South India, showed very unstable thermal stratification, Hence, the above theory does not seem to hold good.

It is clear that there are some factors governing the thermal stratification of North Indian Reservoirs. Both in large and small reservoirs of North India (with some exceptions) indicated distinct and pronounced thermal stratification which is more or less ill defined in peninsular reservoirs.

A further analysis of facts indicate that multipurpose reservoirs of the North India, where water fluctuates appreciably, in general, is prone to such conditions unlike South Indian Reservoirs. Rihand (46, 538 ha), Govind Sagar (16,867 ha), Konar (2589.99 ha), Gitalsud (3,459.50 ha) etc. are the clear examples. Again, in reservoirs large and small in which water does not fluctuate widely *e.g.* Rana Pratap Sagar (20,720 ha), Panchet (7,511 ha), Gandhi Sagar (68,700 ha), Ramgarh (1,260 ha), Maithan (11,491 ha) etc. did not show any thermal stratification, while Mettur Dam (15,346 ha) of South India did not show any thermal stratification even though its water fluctuates widely. This indicates that inspite of some common factors (dead storage and fluctuation in water level in all multipurpose reservoirs) between North and Peninsular reservoirs, some other factors like wind velocity which is influenced by the South West and North East monsoon separately in different regions, location of reservoirs in different latitudes perhaps help in controlling the thermal stratification differently.

Sreenivasana (1978), remarked that although a well defined seasonal thermocline is not characterises of tropical waters, thermal stability or resistance to mixing (RTR) is high in tropical waters, because the density difference per unit of temperature is greater in tropical warm waters than temperate cold waters. He further cited examples saying that in all Indoneosian lake (Ranu Klindungan) for a vertical difference in temperature of 5.2°C (26.2°C–31.4°C) the RTR was 190 g/cm, while it is only 60 g/cm in an Alpine lake for a difference of 7 C (5.2°C–12.2°C) He has further added that in case South Indian waters where vertical different of 3.2° the RTR was 114 g/cm (Aliyar Reservoir). In Bhavani Sagar, for 3.6°C the RTR was 160 g/cm, 135 g/cm in Stanley Reservoir and in Amaravathy reservoir the RTR was 73 g/cm. He further added that in cold-water reservoirs like Pykara and Sandynulla it was low (11.4 to 64.1 g/cm) for depth ranges of 10 to 20 m. He concluded that even small differences in temperature in the tropics, create density gradients which are more stable than thermoclines of temperate lakes.

From Sreenivasan's (1978) observation it is clear that seasonal thermocline is more pronounced in tropical waters not only under high fluctuation of water temperature between surface and bottom, but also under high temperature level (at the surface). That is Indonesian lake 26.2°C–31.4°C) than the lower temperature level

(5,2°C–12.2°C) of Alpine lake. In peninsular reservoirs of India, although the difference between surface and bottom temperature ranged between 7 and 9°C their level mostly varied between 22.4 and 31°C (Table 11) except Pykara and Sandyaulla. It is possibly for this reason their thermal stratification is very rare, ill defined or unstable, while in the North India, under tropical level (upper level 25.5–33.0°C) and subsurface temperature (17.6°–22.3°C) with an average value which is a little lower, the presence of thermocline is more pronounced except in some cases (Tilaiya, Temperature, 24-29 C, maximum depth 12 m and Ramgarh, 18-32°C, depth 16 m), no thermoclines were indicated for which no explanation could be attributed. It is thus clear that temperature levels (surface and bottom) of North Indian Reservoirs are in general showing slightly higher than the Peninsular reservoirs and formation of thermocline in earlier case which generally occur during April and May are prominent and well defined. Getalsud, comparatively a smaller reservoir (3459 ha) and 32.62 m deep (maximum) is a typical reservoir in which thermodine occurred twice a month in some years (Figure 7).

Transparency

Turbidity is caused mainly by heavy particles rather than colloidal silt suspensoids. Clarity of water is generally maintained almost round the year except for a short period during monsoon when the value goes up a little high (205–328 cm). It is interesting to notice that in almost all the larger reservoirs (13200–71963 ha) turbidity is maintained between 20.7 and 169 except in case of Mettur Reservoir where as high as 278 is recorded. The lower level of turbidity however is almost the same both in larger and smaller reservoirs. The wind and wave action during non-monsoon months also cause low turbidity in the entire shallow bays of the reservoir. Turbidity being low and persists for a short duration, penetration of light is easily possible even in deep waters.

Carbon dioxide, Alkalinity and pH

In the process of photosynthesis, chlorophyll bearing phytoplankton utilize the dissolved carbon dioxide. In oligotrophic waters where phytoplankton is poor, the unutilized carbon dioxide is detected when pH remains at a lower level. When free carbon-dioxide is fully utilized in the process of photosynthesis, carbonate alkalinity appears when a rise in pH is noticed. In the productive waters where free carbondioxide is present, it declines during the day when carbonate alkalinity increases. The greater the productivity, the wider is the amplitude of change so that after exhausting the free carbondioxide, bicarbonate carbon is then fixed resulting in decreased methyl orange alkalinity (bicarbonate) and precipitation of carbonate. This leads to a shift to an alkaline pH.

The survey of reservoirs irrespective of thin sizes and location indicates narrow range of fluctuation of pH, which clearly indicates that the water has a high buffering capacity. It is interesting to note that in case of large reservoirs (>10,000 ha) of North India the average pH values ranged between 7.4 and 8.6 and in case of smaller reservoirs (less than 10,000 ha) the values ranged between 7.4 and 7.9. While in case of reservoirs of temperate region such values ranged between 7.5 and 8.5 and 6.9 and

8.1. This means in case of larger reservoirs (>10,000 ha) irrespective of north and temperature region, there is no significant difference in pH value. But in case of smaller ones (Less than 10, 000 ha) their ranges were 7.4 and 7.9, and 6.9 and 8.1, which means the upper level of pH of both the regions always maintained slightly lower value than the larger reservoirs.

Carbondioxide was present in almost all the reservoirs, their concentration in general being well below 10 ppm except on a few occasions when the value rose as high as 24-35 ppm. Carbonate ion was present only in one large reservoir (Ukai of Raj asthan) where the mean value was 9,6 ± 7.53 and carbon diozide 9.5 ± 5.81.

The water always remains alkaline in character due to carbonates and bicarbonates of calcium and magnesium. Their concentration remained low throughout. The lower level attained was 60.6 ppm during monsoon while the upper level estimated was 121.7 ppm. Maharashtra registered lowest value (10-36 ppm) and was followed by North eastern reservoirs (22-32 ppm). In larger reservoirs (above 10,000 ha) on an average lower alkalinity value was 58.83 ppm when the upper level estimated 132.1 ppm and in case of reservoirs between 1000-10,000 ha the average values were 48.11 and 101 ppm. Of the larger reservoirs Mettur Reservoir (15,346 ha) indicated highest value (140–278 ppm) and Rihand Reservoir (46,538 ha) registered lowest value 28-60 ppm. Ramgarh (1260 ha) though a smạll reservoir indicated alkalinity range 129-198 ppm while Haerṇavathy (9163 ha) a comparatively bigger reservoir indicated values between 23-46 ppm only. All the large reservoirs are fairly productive and smaller ones (1000-10,000 ha) stated above are comparatively less productive but are better than other small reservoirs.

Calcium

Wide variation in calcium content and the corresponding correlations and biological productivity have led to classification of lakes based on their calcium content. According to Reid (1961) waters with less than 10 ppm of calcium are designated as poor, between 10-25 ppm as medium and waters with more than 2 5 ppm are highly productive.

Results obtained from various investigations revealed that average lower level of Calcium concentration was 20.1 ppm in the range 2.4 and 54 ppm while their corresponding upper level was 45.24 in the range 16.3 and 93.7 ppm. Highest Calcium concentration 40 and 54 ppm were estimated in Orissa and Madhya Pradesh while their corresponding values of the upper level ranged between 93.7 and 52 ppm. Gujarat also showed high value of Calcium 58 ppm. Among the large reservoirs (10,000–71,963 ha) upper values of calcium was registered in case of Hirakud (52 ppm), Gandhi Sagar (36 ppm, Nagarjuna Sagar (38 ppm) and Govind Sagar (30.4 ppm) and in case of medium reservoirs (1000-10,000 ha) high Calcium content was estimated in Ravisankar (32.44 ppm) followed by Ramgarh 25 ppm.

Specific Conductivity

This is an index of ionizable salts, in other words total soluble solids. Generally in fresh waters this is parallel to bicarbonate and free carbondioxide. The conductivity in productive waters increases from surface to bottom as bicarbonate alkalinity.

The average lower level of electrical conductivity was 147.7 while the corresponding upper level was 680.47 µmhos. 41 and 342 µmhos biong the range of lower level while 57.12 and 1114.7 µmhos is the range of upper level of conductivity value. In larger reservoirs 10,000-66,000 ha) the sp. conductivity value ranged between 185 and 460.2 µmhos, and for reservoirs below 10,000 ha the range was between 117.9 and 263.84 µmhos. Among the larger reservoir highest value recorded was in Nagarjuna Sagar (28,427 ha) (219.6–1114.7 µmhos followed by Tungabhadra (37,814 ha)–240–359 µmhos and Mettur reservoir (15346 ha,)170-350 µmhos). Of the smaller reservoirs (1000–∠10,000 ha) highest value recorded was between 307 and 791 µmhos in case of Ramgarh (1260 ha), followed by Bhavani Sagar (7876 ha) 269.96 µmhos.

Dissolved Oxygen

Dissolved oxygen is a very important critical parameter for aquatic life. Euphotic waters have Klinograde curve with high oxygen content at the surface which drops steeply with depth to a very low level at the bottom. Benthic oxygen consumption and tropholytic action by decomposers make further inroads into the oxygen store. Photosynthesis at the surface and trypholitic activity at the bottom cause klinograde oxygen distribution as is recorded in Bhavani Sagar, Amaravathy, Nagarjuna Sagar, Rihand and Govind Sagar reservoir. In unproductive waters the oxygen curve parallels the temperature curve, since its solution is temperature dependent. Oilgotrophic waters have an orthograde oxygen curve where with depth, the oxygen content varies little.

The survey reveals that the average low level o£ DO varied between 3.5 and 8.0 ppm while the upper level ranged between 7.1 and 14.0 ppm. The difference between higher and lower level of oxygen of the surface water was highest (11.68 ppm) in case of Govind Sagar, followed by Mettur Reservoir 8.1 ppm) and Rihand Reservoir (7.0 ppm). Minimum difference (2 ppm) was also estimated in case of Krishnaraj Sagar. In case of reservoirs between 1000 and 10,000 ha) the difference was very low (2.9-3.7 ppm) Klinograde Oxygen curve of Getalsud (9459), and its thermal stratification, Nagarjuna Sagar (28427), Govind Sagar (16869), Rihand (46538) and Bhavani Sagar (7876) are shown in Figures 7(a, b, c), 10 (b, c) and 12.

Nutrients

Nitrates, phosphates and Silicates are the important limiting factors in aquatic productivity. During the algal bloom both the phosphates and nitrates are impoverished due to their utilization. Silica is correlated with the diatomes and at the time of bloom its content is reduced drastically. The characteristic feature of most of the tropical waters is a low nutrient status with a high turnover rate which results in rapid utilization of the nutrients as soon as they are released by decomposition so that very little remains in water. High range of fluctuation was observed both in case of nitrate (Tr-2.17 ppm) and phosphate (Tr-2.046 ppm).

Nitrate showed low concentration as Trace in Himachal Pradesh and Tamil Nadu; and 0.02 and 0,05 ppm in Gujarat and Orissa, in other cases the concentration ranged between 0.09–1.609 ppm. In larger reservoirs (over 10,000 ha) the nitrate contents ranged between Trace as lower value and 3.2 ppm as higher value, while in case of medium reservoirs (above 1000 ha) the ranges were 0.05 to 2.8 ppm except Ramgarh where trace value has been recorded.

In case of low values of phosphorus, trace value was recorded in the waters of Karnataka (Maharashtra), Himachal Pradesh and Tamil Nadu. In other cases the values ranged between 0.05 and 2,046 ppm. In larger reservoirs trace value (lower level) was observed in almost all the reservoirs while the upper level ranged between 0.04 and 0.12 ppm.

The average concentration of Silica ranged between 6.62 and 12.75 ppm. In general larger reservoirs maintained a range between 7 and 19.9 ppm except Nagarjuna Sagar where the upper level was fairly high (40 ppm) while medium size reservoirs (above 1000 ha) indicated a level of 1.3 to 18.46 ppm of silica content, with an exception of Aaravatty (850 ha) where the concentration rose as high as 38.5 ppm.

Chapter 12
Biological Status

Ecology of reservoirs is somewhat different from the rivers–they are neither lotic nor lentic but a mixture of both. Normally a large volume of water is kept stored as dead storage which is normally maintained and not allowed to flow except on special occasion. Over this dead storage level large volume of water is allowed to be stored up to a particular level during monsoon and depending upon the requirement (irrigation, power generation and water supply to industries and cities etc.) the water is allowed to be discharged from this stored reservoir. As a result in the absence of adequate time for its stability the littoral and sub-littoral jones gradually get exposed and fail to contribute productivity of the reservoir water effectively. Although the river feed the reservoir round the year, its supply gradually is reduced till it reaches its minimum around June/July until monsoon breaks. During late monsoon or draught period the reservoir indicates a different picture. Under such stressed condition biotic communities are to grow and reproduce to help for secondary and tartiary production.

Normally river water feeding the reservoir brings sufficient concentration of plankton which depends upon the morpho-ecological condition of the reservoir grows maintaining their own qualitative and quantitative character. Plankton studies were made in different reservoirs, large, medium and small spread over the country, but unfortunately their mode of representation being different in a number of reservoirs *i.e.* unit/litre; ml/m^2; unit/m^2; ml/m^3; mg/m^3 etc. This has become difficult to show a better comparative picture among them. In some cases concentration of individual species are shown in percentage without mentioning their concentration. However, an indication of concentration of plankton in some reservoirs of different categories are available which indicates large reservoirs like Hirakud (71963 ha) (Orrissa), Gandhi Sagar (66000 ha) (M.P.), Pong (H.P.) and Govind Sagar (16367 ha) (H.P.) indicate their range of concentration as 2590-42474 u/l, 3211 u/l (Max.); 669-12786 u/l; and 249-1382 u/l (1989-90). Tungbhadra (37814 ha) (Karnataka) although a large reservoir showed its productive period 85-233 u/1 only. In case of smaller reservoirs like Konar 2590 ha (Bihar), Ravisankar Sagar 9540 ha (M.P.) and Ramgarh 1963 ha (Rajasthan) their concentrations are found to be 3570 u/1 (highest density), 534 u/1 (average) and 2890-14680 u/1. Such concentrations are definitely encouraging. The study shows that well developed planklen cycles become established immediately after filling up of the reservoir. A pattern of seasonal pulses occur generally during February–June and the other between October and December.

Jhingran (1988) has recorded some interesting observations regarding dominance of phytoplankton over zooplankton which vary from year to year and by and large the predominant group was blue greens, a group not noted for passing their high productive capability directly up the tropic web. Where as myxophyceae were observed to be mainly represented (Microcystis) in reservoirs of lower latitude, the same were replaced by dinoflagellates (ceratiurn) in reservoirs located in higher latitude. Bacillarioplyceae were mainly constituted by *Synedra, Navicula, Flagilaria, Melosira* and *Gyrosigma, Chlorophyceae contained–oedogonium, Spirogyra, Pediastrum, Botryoccous, Pandorine* and *Eudorina* as dominant species forms.

Copapods, the main constituents of zooplankton were mainly represented by *Diaptomus* and *Cyclops*; Rotefers by *Keratella, Brachionus, Polyarthra* and *Filinia*.

The general trends of absence were found to change due to rainfall and nutrients from run off. Maximum numbers developed from April to June and minimum July–August in most of the reservoirs. Seasonal cycles of rotifers have been found to follow that of Phytoplankton closely (Benson, 1968).

Productive waters of Gangetic plain, Deccan plateau. South Tamil and Orissa invariably influence the standing crop of *Microcyrtis*. (Sugunan, 1995) recorded spicies is almost omnipresent in Southern Peninsula except in reservoirs of Karnataka and Kerala which indicate this oligotrophic character resulting comparatively poor plankton concentration with desmids and other green algae. Sugunan (1995) also observed that due to scanty rainfall and poor flushing rate favours microcystis. The desmid dominated plankton is the distinct character of the North Eastern Reservoirs. It is interesting to note that consequent transformation of lotic environment into lentic system saprophobes disappear from the water and allow rapid multiplication of saproxeus. Microcystis finding a favourable condition with the new environment burst into blooms and dominate. In may reservoirs orientation of fluviatile plankton are clearly indicated in the composition of plankton in lotic, centric and cove zones.

The fluviatile lotic sector although indicate a lower plankton density often shows better diversity and evenness indices as compared to lentic and bay zones, the still water of which are characterised by higher concentration of dominance and low evenness.

Preimpoundment vegetation and associated debris offer varied nutrient and substrate relations. Decomposition of these materials combined with erosion–redeposition processes along with the newly formed shore line, produces rapid changes in the bottom biotope. Benthos formation progresses in a highly unstable condition of the reservoir water due to its release particularly during the dry months for the purpose of irrigation, power generation and water supply in industries and cities. As a result almost in all the reservoirs (especially multipurpose reservoirs) whose shore line development index ranging from 1.6 to 12.86 gradually become exposed. Available data from the random sampling of reservoirs of different categories large, medium and small, spread over the country indicate range of concentration of macrobenthos as 179.7-385.5u/m^2 in Tungabhadra (37814 ha), 2500 u/m^2 (chironomids) in Nagarjuna Sagar (28427 ha), 286-571 u/m^2 in Pong Reservoir (24539 ha) and in Rihand (46538 ha) Reservoir it is only 32-70 u/m^2; while in the medium size reservoir like Govind Sagar (16867 ha) the average concentration was 466.5 u/m^2, and in Krishnaraj Sagar (13200 ha) the average concentration was 933 u/m^2 and in smaller reservoir Ravisankar Sagar (9540 ha), the range was 343-600 u/m^2, Haemavaty (9162 ha) 2042-4174 u/m^2, Bhavani Sagar (7876 ha), the standing crop was 1194/m^2 which increased with depth. Further in smaller reservoirs (140-646 ha) the organisms showed a very high value, 1884-3828 u/m^2. The organisms grew well in littoral and sublittoral zones around 10 m. depth which in some cases extended to 40-50 m. The number of organisms could have further increased, but the purpose is defeated when draw down of water from the reservoir is considerable as stated earlier. As a result large littoral and sublittoral area get exposed. The author has observed that millions of small dead shelled organisms (Bivalves and Gastropods) are congregated like a necklace along the margine of water on the rocky bed of Tungabhadra Reservoir during 1964-65. This is found to be an annual phenomenon. Thus a large number of organisms get destroyed in reservoirs and the rest are brought down to the deeper water with the help of wind action associated with the draw down of water. Since fluctuation of water level in smaller reservoir is much less compared to larger reservoirs, benthos can remain almost undisturbed and as a result their abundance is much higher than the larger reservoirs. Bachhra, Bagla, Loni, Tirumurthy and Aliyan are such examples.

However, reservoirs of various sizes, irrespective of their sizes, show the pattern of benthic communities more or less similar with some variations. A list of diversity of organisms generally encountered are indicated below:

Table 15: List of Benthic Communities Encountered in Different Reservoirs

1.	Dipteran larvae	2.	Caddis worm
3.	May fly nymph	4.	Dragonfly nymph
5.	Cligochaete	6.	Bivalve shells
7.	Insects (Ephemicoptra)	8.	Chironomids
9.	Molluses (Gastropods)	10.	Melania
11.	Corbicula	12.	Chaoborus larvae
13.	Worms	14.	Crustacea
15.	Small fish	16.	Tad poles

Periphyton takes the form of a brown of green layer at the surface of submerged objects and form food source of browsing fishes. This group is reported mainly by diatoms, green algae and blue greens. Pheriphyton develop best in litteral areas of the reservoir. In Konar Reservoir as reported by Jhingran (1988) periphyton group was constituted of an algal complex comprising green algae (*Mougeotia, Hormidium, Oedogonium,* blue green algae (*Anabaena, Phormidium, Lunghys*) along with diatoms (*Navecula, Surirella, Synedra, Fragilaria, Gyrosigraa*).

Chapter 13

Transformation of Energy in Tropical and Subtropical Impoundments

Energy is the capacity of doing work and all living organisms require energy in one form or the other. The biotic communities (producers, consumers and decomposers) in an ecosystem are linked with one another with energy chains, unidirectional flow of energy and nutrient cycles are the two great principles of general ecology. Hence, apart from different factors responsible for the productivity of reservoirs discussed earlier, a complete knowledge of the interrelation ships among organisms, flow of energy and nutrients from one level to the other and the role of environmental parameters in the energy transformation processes is of great importance for the correct understanding of the ecosystem and better utilisation of energy source.

No detailed study has been made on the energetic aspects of man made impoundments in India which are the store houses of a large number of organisms with various energy chains. Ganapati (1970), Sreenivasana (1972), Ganapati and Sreenivasana (1972) and Natarajan and Pathak (1980) are the limited few who studied the bioenergetics of some man-made lakes. Lack of understanding of ecological principles of energy transformations and improper management practices have resulted in very poor energy harvest (10 kg of fish/ha or approximately 12×10^7 cal/ha of energy on an average basis) Natarajan and Pathak (1983). The details of energy status studied so far are highlighted under three important aspects as:

1. Influence of latitude on the availability of light energy in reservoir water and its transformation to chemical energy by photosynthetic organisms.
2. Productivity (in terms of energy) in reservoirs from available radient energy and finally.
3. Energy transformation and fish yield.

(*i*) Influence of latitude in five reservoirs of different sizes situated in different regions indicate their respective geographical locations, their available light energy with their corresponding transformation to chemical energy by photosynthetic organisms are indicated below in Tables 16A and B.

Table 16A: Influence of Latitude on the Variations in the Availability of Energy in Reservoir Water

Reservoir	*Area at FRL*	*Location*	*Availability of Light Energy cal m^{-2} yr^{-1}*	*Available Chemical Energy cal m^{-2} yr^{-1}*	*Transformation of Photo-synthetic/ utilization of Light Energy*
Bhavani Sagar (Kerala)	7285	11°25'	213 x 10^4	8781	0.41 per cent
Nagarjuna Sagar (A.P.)	28474	16°4'	205 x 10^4	5959	0.29 per cent
Rihand (U.P.)	46538	24°	188 x 10^4	3970	0.2 per cent
Ramgarh (Rajasthan)	1265	27°12'	183 x 10^4	8236	0.49 per cent
Govind Sagar (H.P.)	16867	31°25'	172 x 10^4	11696	0.68 per cent

Source: Jhingran A.G. (1990).

It is interesting to note that by their very location, the radiations reaching these waters vary considerably. For example Govind Sagar located at 31°25' shows lowest quantum of solar energy (172 x 10^4) followed by Ramgarh (27°12'), Rihand (24°), Nagarjuna Sagar (16°4) and Bhavani Sagar (11°25') when their corresponding quantum of available light energy were 183 x 10^4, 188 x 10^4, 205 x 10^4 and 213 x 10^4 respectively. The transformation of chemical energy from available light energy by photosynthetic organisms were however in the order 11696 cal^{-2} yr^{-1} or 0.68 per cent for Govindsagar, 8236 cal m^{-2} yr^{-1} or 0.49 per cent for Ramgarh, 3970 cal m^{-2} yr^{-1} or 0.20 per cent for Rihand, 5959 cal m^{-2} yr^{-1} or 0.29 per cent for Nagarjuna Sagar and 878 cal m^{-2} yr^{-1} or 0.41 per cent for Bhavani Sagar. The variation in the quantum of chemical energy is often the qualitative and quantitative abundance of the producer communities that determines the photosynthetic efficiencies rather than the actual amount of available solar energy.

(*ii*) Productivity (in terms of energy) in some reservoirs from the available redient energy is also estimated earlier by Sreenivasana (1969) in some smaller peninsular reservoirs, and are compared to some larger reservoirs stated earlier, studied by Natarajan and Pathak (1983). Comparative productivity (in terms of energy) in some reservoirs located in different latitudes are indicated in Table 16B.

A comparison of the photosynthetic energy with the energy harvest as fish shows an interesting pattern both in larger and smaller reservoirs. A good conversion indicates the efficiency harvest of fish and of manage ment practices. Nataranjan and Pathak (1983) in their study observed that the two tropical impoundments Bhavani Sagar and Nagarjuna Sagar receive almost similar light radiation and have quite high photosynthetic efficiency but the energy harvest as fish is much higher in

Table 16B: Comparative Productivity (in terms of energy) Fixation and Conservation in Some Reservoirs Located in Different Latitudes

	Source: Natarajan A. V. and V. Pathak (1983)				*Source: Sreenivasana, A (1972)*					
Productivity	*Govind Sagar (16,867 (FRL)*	*Rihad (46,538) (FRL)*	*Nagarjuna Sagar (28,427) (FRL)*	*Bhavani Sagar (7,875) (FRL)*	*Bhavani Sagar (7,875)*	*Stanlay (15,346)*	*Santanur (2,010)*	*Krishnagiri (1,448)*	*Amaravatty (850)*	*Sandynulla (258)*
Location	31°35'N	24°N	16°34'N	11°5'N	Location will be around 11°5'N (being in the same state)					
Total Radient Energy cal/m²/yr x 10^5	6278	6877	7483	7775	7775	7775	7738	7720	7847	8578
Photosynthetic Oxygen Production cal/m²/yr x 10^6	4.269	1.387	2.175	3.205	8.162	8.147	8.100	5.610	11.320	8.121
Photosynthetic efficiency per cent	0.682	0.202	0.290	0.412	1.050	1.047	1.047	0.727	1.442	0.947
Fish Production										
(i) g/m²/yr.	5.36	0.44	0.79	5.94	2.78	6.29	2.70	2.05	16.00	4.54
(ii) Cal/m²/yr.	6432	528	948	7128	3336	7550	9241	2460	19,230	5454
Conversion efficiency										
(i) Photosynthesis to fish (per cent)	0.151	0.038	0.043	0.220	0.0649	0.927	0.1141	0.1449	0.1698	0.0671
(ii) Light energy to fish (per cent)	0.0010	0.000076	0.00013	0.00091	0.00191	0.000971	0.00194	0.000319	0.002451	0.000636

* Procedures followed for estimation of various factors (energy) are given by Natarajan and Pathak (1983).

Bhavanisagar than Nagarjuna Sagar. The conversion of photosynthetic energy fixation to fish and light to fish is 0,22 per cent and 0,0009 per cent respectively in Bhavani Sagar but in Nagarjuna Sagar the values are 0.043 per cent and 0.00013 per cent only.This clearly indicates that the management has failed in Nagarjuna Sagar and.the energy harvest is much less than expected though the limnological features of the two reservoirs shoxtf their productive nature. Govind Sagar, another productive reservoir from limnological point of view also shows high values of conversion from chemical energy fixed to energy harvest (0.15 per cent) or light to fish (0.001 per cent). On the other hand of the six reservoirs studied by Sreenivasana (1969), Bhavani Sagar, Stanley and sandynulla with almost similar light radiation show quite high photosynthetic energy to fish (0.0649 per cent, 0.0927 per cent and 0.0671 per cent) which are even better than Rihand and Nagarjuna Sagar; and light to fish (0.00191 per cent, 0.000971 per cent and 0.000636 per cent). In case of Sandynulla the energy harvest is much less than expected. In case of other reservoirs, santanur, Krishnagiri and Amaravathy although the photosynthetic efficiency is high (0.1141 per cent, 0.1449 per cent and 0.1698 per cent) but their corresponding light energy to fish were 0.00194 per cent, 0.000319 per cent and 0.002451 per cent respectively which shows that the management is not up to the mark but better than larger reservoirs., Thus the overall picture is that smaller reservoirs are better managed and is reflected from better energy harvest. Thus the studies of conversion efficiencies, light to chemical (photosynthesis), chemical to fish and light to fish clearly indicate the extent of utilisation of the energy resource and the management measures to be adopted for mixing the energy output from man made ecosystem. The low energy output from various reservoirs in the country is due to lack of understanding of the energy relationship at various levels and the improper management practices resulting in partial utilization of the energy resource.

(*iii*) Four reservoirs located in different latitudes differ considerably in the energy transformation from light to chemical and its further utilisation shown in Tables 17 and 18 are highlighted by Natarajan and Pathak (1983).

In Bhavani Sagar about 68.6 per cent of the energy output is contributed by primary consumers, 3.4 per cent by secondary and 28 per cent by tartiary consumer. Of the 68.6 per cent, only 5.4 per cent is through grazing chain and 63.2 per cent through detritus chain. The bottom energy in the reservoir is considerably high (9,53,00 x 10^6 Cal/ha) and as the major flow of the reservoir is through detritus chain, the reservoir shows high energy of conversion from light to fish or chemical to fish and correspondingly better energy harvest.

Nagarjuna Sagar on the other hand had only 26.4 per cent, of the energy output from this reservoir is contributed by primary consumers, 12.5 per cent through grazing and 13.9 per cent through detritus chains respectively, and the remaining 73.6 per cent by secondary (34.5 per cent) and tartiary (39.1 per cent) consumers. The reservoir has very high energy resource at the bottom (14,00,000 x 10^6 Cal/ha) but this is not entirely utilised through primary consumption. In this reservoir *Pangasius pangasius* is the dominating catfish and it utilises bottom energy through secondary consumption contributing 34.5 per cent of the energy obtained. The non-utilisation of

Table 17: Energy Status in Different Reservoirs

Reservoirs	*Incident Light Energy Cal/ha/yr x 10^6*	*Available Energy*		*Energy Utilization*		*Secondary Consumption Cal/ha/yr x 10^6*	*Tertiary Consumption Cal/ha/yr x 10^6*
		Chemical Energy Fixed by Producers Cal/ha/yr x 10^6	*Bottom Energy Cal/ha/yr x 10^6*	*Through Grazing Chain Cal/ha/yr x 10^6*	*Through detritus Chain Cal/ha/yr x 10^6*		
Bhavani Sagar	77,75,000	32,050	9,53,000	3.60	43.80	2.40	19.30
Nagarjuna Sagar	74,83,000	21,7000	14,00,000	1.05	1.10	2.90	3.59
Rihand	68,77,000	13,900	3,60,000	3.8	–	1.3	–
Govind Sagar	62,78,000	42,700	30,70,000	23.00	16.90	24.70	0.80
(Efficiency of Conversion (per cent)				*Per cent Utilization of Energy*			
Bhavani Sagar		0.412	5.4	68.6	63.2	3.4 per cent	28 per cent
Nagarjuna Sagar		0.290	12.5	26.4	13.9	34.5 per cent	39.1 per cent
Rihand		0.202	80.0	-	-	10-15 per cent	–
Govind Sagar		0.682	35.1	60.1	25.0	37.8 per cent	2.1 per cent

Source: Natarajan A.V. and V. Pathak (1982).

the vast energy at the bottom specially through primary consumption has led to the poor harvest of energy from this reservoir, thus demanding application of suitable management to enhance the output of energy.

Table 18: Important Fisheries of Different Reservoirs and their Energy Contribution

Species	*Average yield kg/ha/yr.*	*Cal/ha/yr x* 10^6
	Nagarjuna Sagar	
L. fimbriatus	0.62	0.98
L. rohita	0.06	0.07
L. calbasu	0.67	0.80
C. mrigal T.	0.09	0.11
T. Khudree	0,17	0.20
C. catla	0.16	0.19
P. pangasius	2.26	2.71
M. seenghala	0.54	0.65
M. aor	1.20	1.44
S. childreni	1.05	1.26
W. attu	0.20	0.24
	Govind Sagar	
L. rohita	10.0	12.0
L. dero	8,7	10.4
H. molitrix	0.5	0 6
C. mrigala	3.7	4.4
C. carpio	10.4	12.5
C. Catla	13.5	16.2
T. *putitora*	7.1	8.5
M. seenghala	0.8	0.9
	Bhavani Sagar	
L. rohita	3.5	4.2
L. finrbriatus	0.1	0.12
C. mrigala	7.2	8.6
P. dubius	2.6	3.1
L. calbasu	26.8	32.1
C. catla	2.4	2.9
M. aor	12.5	15.0
W. attu	6.8	8.2
	Rihand	
C. catla	3.97	4.68
Others	0.43	0.62

Source: Natarajan, A.V. and V. Pathak, 1983

In case of Govind Sagar 60 per cent of the energy output is contributed by primary consumers of which 35.1 per cent is through grazing and 25 per cent through detritus chain. The remaining 39.9 per cent of the energy is obtained mainly as secondary consumers. Owing to the proper utilisation of energy by the existing fishes this reservoir shows quite high conversion ratio either from light to fish or chemical to fish. However, both the bottom energy (30,70,000 x 10^6 Cal/ha) and autotrophic chemical energy (42,700 x 10^6 Cal/ha/yr being very high there is still enough scope to enhance the energy harvest from this reservoir.

Finally in case of Rihand reservoir 90 to 95 per cent of the energy is utilised by a single species C. *Catla* owing to the single species and single size group dominance, this reservoir has shown minimum energy output. Further, this reservoir has minimum energy reserve at the bottom compared to other reservoirs. Thus the level of variation is attributed to the quantitative and qualitative variation in the species spectrum of the lakes. Two main pathways through which primary chemical energy finds its way to fish are the grazing chain and the detritus chain. The contribution by both the pathways to the total availability of energy is required for deriving the species combination most suited to the ecosystem, Thus once the energy pathway is established, right management can ensure a higher rate of energy harvest in the form of fish.

Chapter 14
Reservoir Pollution

Except some, reservoirs have come up mostly from fiftys onwards. Like rivers, reservoirs are also found to be affected due to industrial development and urbanisation in the country. Besides direct entry of compex industrial and municipal wastes, they are also carried by upstream rivers which joins the main reservoir along with the inflow;

Not much of work has been done on pollution problems in Indian reservoirs. Of late Joshi (1990) has reviewed the various pollution problems done so far in India. On the contary the author has classified ecodegradation of reservoirs from thermal power plants located near the reservoir and the pollution load from various complex wastewater either directly discharged or carried by the incoming stream. Broadly these are from chemical which include heavy metals and pesticides, and thermal power plants/coal washeries (Table 19).

Table 19: Reservoirs Facing Pollution Problems (Joshi, H.C., 1990)

Sl.No.	Reservoirs	Name or River		Source of Pollution
1.	Rihand (46,620 ha)	(U.P.)	Rihand	Chemicals.
2.	Panchet Reservoir	(Bihar)	Barakar	0.42 million m^3/d of Industrial Wastes from 31 industry. Pollutants–Salt, Phenol, Cyanide, oil/grease creating a BOD of 2022t, 3171t COD. Heavy metals Zn, Cu, Cr. Cd and lead, fly ash.
3.	Bhavani Sagar (786/Ra)	(Tamil Nadu)	Bhavani	Viscose Rayon factory effluents.
4.	Bhairmangala Res.	–	–	Minicipal and Trade effluents of Bangalore City, Heavy metals–Zn, Cu, Cd, Cr, P6 and Hg.

Contd...

Table 19–Contd...

Sl.No.	Reservoirs		Name or River	Source of Pollution
5.	Getalsud (34515 ha)	(Bihar)	Subarnarekha	Heavy engineering chemicals and sewage.
6.	Tungabhadra (37,814 ha)	(Karnataka)	Tungabhadra	Paper, Iron and Steel, rayon, Chemicals and Sewage, Heavy metlas.
7.	Hussain Sagar	(A.P.)	Musa	Trade effluents, sewage from Hyderabad city.
8.	Gandhisagar (68,700 ha)	(H.P.)	Chambal	Textile, Chemicals trade effluents from Indore, Ujjain and Kota.
9.	Hirakud	(Orissa)	Mahanadi	Paper Mill.
10.	G.B.Pantsagar	(U.P.)	Rend	Thermal power plant, Coal washing, chemicals.
11.	Sandynulla	(Tamil Nadu)		City Sewage.
12.	Dal Lake	(Kashmir)	Jhelam	Toxic metals (Sewage and Waste)

The nature of pollution studied so far in some of the reservoirs large and small are indicated below.

Rihand

The characteristics of Kanoria chemical disposing into the reservoir are:

pH: 9.2

Total alkalinity: 4770 mg/l.

Sp. conductivity: 12816 μ mhos.

Chlorides: 51,735 mg/l.

Free Chlorine: 1224 mg/l.

Effect

1. Severe toxic effect on phyto and zoo plankton reported and direct fish kill reported due to high clorine (Arora *et al.*, 1970).
2. Monitoring of heavy metals in the reservoir has shown significant level of mercury in water, fish and sediments. Agrawal and Kumar (1978) have also reported the presence of mercury in the effluent of soda Manufacturing factory, discharging its effluent into the Rihand reservoir.
3. Panwar *et al.* (1979) recorded heavy fish mortality in Summer of 1977. The dead fish comprised *C. Catla* (2.5–3.0 kg), *M. aor* (0.3–0.6 kg) and *C. mrigala* (0.35–1.50 kg) and *C. garua* (72–85 kg). Heavy mortality of carp and cat fish fingerlings were also recorded during monsoon months. Absence of fish food organisms, plankton and bottom biota in the polluted zone confirms that the effluents which are highly toxic in nature are effecting the biomass in the reservoir.

4. Recent observations made by Joshi (1990) in Rihand Reservoir has shown wide distribution of pesticide residues in water, sediments, plankton and fish. The residues of different isomers of BHC and DDT and its metabolites (DDE, DDD) were found to be present invariably in fish and plankton. A chromatogram of mussel tissues of the fish *Catla catla* is illustrated in Figure 1. This shows the significant present of αBHC, γBHC, βBHC and δBHC all isomers of benzene hexachloride and op′ DDE, pp′ DDE, op′ DDD, pp′ DDD, OPTDDT and pp′ DDT, all isomers and metabolites of DDT. Since, the reservoir is situated in a relatively secluded region where industrial and agricultural activity had been very low, these observations assume immense importance.

Besides, thermal power plants located around the reservoir are also the source of pollution. Chandra *et al.* (1985) have reported heavy fish mortality in this reservoir due to thermal discharges from one such plant.

Panchet Reservoir

Gopalakrishnan *et al.* (1966) recorded adverse effect of effluents from coal washeries on the recruitment of Indian major corps. Besides, wastes from Sindri Fertilizer factory are reported to kill spawn in large quantities in Panchet Dam [Konar *et al.* (1990)].

Banerjee and Mukherjee (1990) has made an elaborate study of Panchet Reservoir with regard to pollution problems in this reservoir. The reservoir has highly strained environment receiving complicated wastes of about 0.42 mm^3/day of industrial effluents from 81 industries. The effluent admixture from different industries has been estimated to contribute solids, cyanide and oil/grease creating a demand of 2022 t BOD_5 and 31711 t of COD/day. High concentration of heavy metals like Zinc, Copper Chromium, Cadmium and lead have also been estimated. The fly ash from thermal power Station has reduced the euphotic zone and zeta potential of the entire ecosystem to a considerable level. The continuous influx of toxic effluents has adversely affected the food chain of Indian major Corps resulting in the dominance of resistant species like *Mystus aor, Notopterns notopterus* etc. The fish output is only 2.7 kg/ha/yr.

Bhavanisagar Reservoir

Impact of viscose industry effluent was confined to a certain stretch below the river only when the inflow was very low pH recorded was below 4, D.O. 1.0 ppm. with saturation percentage 14.0 at Siramughai. Dead plankton and fishes were observed on the surface of water, The discharge of acidic effluents and subsequent deplition of DO and low pH were the Cause of mortality of plankton and fishes. Such effluent did not show any adverse effect in the main reservoir (Final Report 1971-81 All India coordinated Res. Proj. IFRI Series 1, May, 1981).

Byramangala Reservoir

Joshi (1990) made an extensive study in this reservoir receiving municipal and Trade effluents of the Bangalore City, Such wastes have shown appreciable accumulation of metals like–2n, Cu, Cd, Pb and Hg in sediments and plants. The results presented in Tables 20 and 21 show that metals are accumulated in plants and soil by more than 350 to 4 500 times of the metal level found in water. Results also indicate that the adsorption on soil and the bio-adsorption by plant is nearly the same in case of Zn, Cu and Hg. While bio-absorption of Cd and Pb by plants is less than the absorption by the sediments. Thus it signifies that while Zn, Cu and Hg are continuously removed by the plants, Cd and Fb are gradually being accumulated in the bottom soil.

Table 20: Status of Heavy Metals in Byromangala Reservoir

Metals	*Concentration*		
	Water (μg/l)	*Sediments (μg/g)*	*Plant μg/g (dry wt)*
Zinc	87–130	50–197	76.5–207.8
Cu	28–52	38–64	33–143
Cd	nd–15	32–106	0.4–2.1
Cr	nd–	0.88–1.32	0.42–0.7
Pb	16.22	53.4–101.2	5.3–9.0
Hg	0.08–8.12	0.14–0.4	0.29–0.63

Table 21: Accumulation of Heavy Metals in Sediments and Plants (*E. Crassipes*)

Metals	*Rate of Accumulation (Water : Soil/Plant)*	
	Sediment	*Plant*
Zn	1 : 1400	1 : 1300
Cu	1 : 1100	1 : 1600
Cd	1 : 8000	1 : 350
Pb	1 : 3500	1 : 370
Hg	1 : 3000	1 : 4500

Gitalsud Reservoir

The study made in the upstream of Gitalsud Reservoir to determine the possible adverse effects of Heavy Engineering corporation factory effluents and town sewage on water quality and plankton.

The effluents contain mainly oil, phenol and coal tar. The study suggests that water quality gets much improve before R. Subarnarekha joins the Gitalsud Reservoir. (all India coordinated Res. Project, Final Report 1971-81) IFRI–Series 4, March, 1984.

Tungabhadra

Occurrence of large scale fish mortality in the Tungabhadra river by Harihar Polyfibres and Grasim Industries are as follow:

			Remarks
1.	First large scale death of fish occurred 2 yrs. back *i.e.* 1982	("*The Statesman*" Kolkata) Oct. 15, 1984)	Several tractors–trollies loaded with dead fish were brought by the fishermen to the District head quarters who intern seized the fish immediately to prevent their being sold in the market.
2.	Between March 24–29 and in May 1984.		Based on the results of investigation carried out by various regional and national institutions, including those of the University who unanimously pointed towards the two Industrial units as being the source of pollution which was causing the fish to die is reported by Dr. Rame Gowda, V.C. Karnataka University.
3.	February 14, 1984 (Anonymous/1984)		

As a squel to large scale fish mortality in the river Tungabhadra on the, 14 February, 1984 (Anonymous, 1984; Joshi and Sukumaran, 1991) made a short term study during September 1985 to June 1986 in the river Tungabhadra near Harihar town (Karnataka State) receiving 44000 m^3/d of effluents (33000 and 11000 m^3) from Harihar Polyfibres and Grasim Industries and discharge it into the river within a 200 m stretch. Their findings are discussed below:

The impact of effluents on the water quality was discernible from the total absence of dissolved'oxygen near the Rayon outfall during the lean flow periods in the river. The brownish Coffee colour of the Pulp effluent was observed till Airani, 10 km, down stream of the two out falls. The alkalinity 140 mg/l and hardness 144 mg/l of the unpolluted stream waters increased considerably to 397 and 465 mg/l near the out fall and the impact continued till Airani when the other bank of the river remained unaffected by the effluents. The average zinc level at the Rayon outfall was 3.2 mg/l. The highest value (4.85 mg/l) was recorded in February 1986. The value however decreased to 0.1 mg/l at 55 km, down stream. Sediment and fish from down stream locations showed high zinc levels. Appreciable value of mercury in water (0.5 µg/l) and sediment (0.26 µg/g) near the pulp outfall indicated occasional discharge of mercury in the river. Average zinc and mercury levels in the muscles, liver and kidney of fish were higher in the fishes collected from below outfall 2, as compared to those collected from above outfall and below outfall 3. Accumulation of zinc was maximum in Kidneys, followed by liver and muscles (Table 22).

Hussain Sagar

The reservoir got polluted due to indiscriminate release of industrial effluents and organic sewage into the lake. As a result fish fauna changed considerably and at present only hardy species dominate in the lake. A comparison of the biology of the common species *Channa punctatus*, *Heteropneustes fossilis* and *Mystus vittatus* from the polluted lake with those of other oligotrophic waters has shown significantly low growth rate and fecundity, high parasitic infection and gut content containing mostly insects, insect larvae, worms etc. Feeding activity is also found to be less compared to that of the specimens from oligotropic waters (Babu Rao, 1990).

Table 22: Heavy Metals (Zn and Hg) in Fishes Collected from Different Locations in the Tungabhadra River Near Harihar

Fish (Nos.)	*Zinc (µg/g Wet wt.)*			*Mercury (µg/g Wet wt.)*		
	Muscle	*Liver*	*Kidney*	*Muscle*	*Liver*	*Kidney*
		At Nadiahamahalli				
Puntius pulchellus (5)	18.2	72.0	118.8	–	–	–
Chrrhina reba (5)	16.1	77.5	151.0	–	–	–
Puntius buckeli (5)	18.1	89.7	142.8	–	–	–
Channa marulius	16.0	66.0	173.9	0.18	0.74	0.71
		At Harihar AOF				
Puntius pulchellus (2)	4.6	–	–	–	–	–
Puntius Kolus (2)	5.6	–	–	nd	–	–
		At Airani BOF3				
Puntius pulchellus (2)	6.8	–	–	0.056	–	–
Puntius Kolus	10.0	–	–	0.020	–	–

Source: Joshi, H.C. and P.K. Sukumaran (1991).

Dal Lake

There is alarming news about the world famous Dal Lake situated in the heart of Srinagar city. The scenic lake has already reduced to 12 sq.km. from 24 sq.km. in 1962. It has excessively high levels of toxic metals owing to sewage and waste. According to comptroller and Auditor General (CAG) released recently the lake water shows high percentage of metals like iron, manganese, copper, lead, cadmium and arsenic. Their concentrations recorded as–Iron–68,57, manganese–5.89, copper–4.89 mg/l while lead, cadmium, chromium and arsenic persisted in concentrations–3.37, 0.10, 0.81 and 48 mg/l against their respective permissible limits as 0.30, 0.10, 0.5, 0.1, 0.05 and 0.05 mg/l respectively. These heavy metals concentrate in the living tissue through food chain which results in high level of these elements in aquatic life that are ultimately being consumed by human beings. Pathological effects of these elements can cause damage to brain, lever and kidneys. Due to destruction of breeding grounds and environmental stress fish species have undergone decrease in size and population.

Chapter 15
Fish Population and Yield

Despite the cataclysmic faunistic changes associated with the impoundment, Indian Reservoirs preserve a rich variety of fish species. The ichthyofauna of a reservoir basically represents, the faunal diversity of the parent river system. On the basis of studies conducted so far as stated *Sugunan* (1995), large reservoirs on an average harbour 60 (sixty) species of fishes of which at least 40 (forty) contribute to the commercial fishery. The fast growing Indo-Gangetic carps, popularly known as Indian major carps occupy a prominent place among commercially important fishes. More recently, number of exotic species have contributed substantially to commercial fishes.

Jhingran (1988) options that fisheries of Indian Reservoirs constituted indigenous and stocked fish population. Among the former, the Gangetic major carps occupy a prominent place in the North Indian Reservoirs both as naturally occurring and stocked fish. In addition to these, they also harbour *Labeo bata, Punitus sarana, P. Chaounio* and *Cirrhinus reba*. However, in Gandhi Sagar (M.P.), the major carps constituted very low percentage of the total fisheries as the dominant fisheries comprised *Tor putitora, Labeo dero* and *Schizothorax plegiostomus*.

In peninsular reservoirs, the indigenous fishes forming the commercial fisheries are *Cirrhinus cirrhosa, C. reba, Labeo kuntius, L, firfibriatus, Puntius dubius, P. sarana, P. Carnaticus, P. Kolus, P. dobsonj, P. hexagonolepis, Tor tor, Thunnichthys, sandkhol*, and *Osteobrama vigorsii*. In addition to these species Gangetic major corps are also stocked.

Among large catfishes the Indian Reservoirs contain *wallage attu, Aorichthys seenghala, A. aor, Silonia silondia* and *Pangasius pangasius, Silonia childrenil and Pseudotropius takree* have established in Tungabhadra reservoir (David, 1969) and *Mystus vittatus* in Poondi reservoir (Chacko, 1969).

Indian reservoirs also harbour a sizable population of trash fishes like *Ambasis nama, Esomus danrica, Aspidoparia mora; Amblypharyngodon mola, Puntius sophore, P. ticto, Oxygaster bacaila, Lanbuca lanbuca, Barilius barila, Osteobrama cotio* and *Gudusia Chapra*. Most of these trash fishes compete for food with Indian major corps tending to reduce the overall productivity of the reservoir. Broad categorization of species encountered and the effect of reservoir formation on fishery are shown in Tables 23 and 24 A and B.

Table 23: Broad Categorization of Species, Encounted in Indian Reservoir

1. The Indian Major Cors: *Labeo rohita, L. Calbasu, L. fimbriatus Cirrhinus mrigala, Catla catla.*
2. The Mahseers: *Tor tor, T. puttitora, T. Khudree, Acrossochilus hexagonolepis.*
3. The minor carps including snow trout and peninsular corps: *Cirrhinus cirrhosa, C.reba, Labeo kontius, L. bata, puntius sarana, P. dubius P.Carnaticus, P.Kolus, P.dobsoni, P. chagunto, Schizothorax plagiostomus, Thynnichthyes, sandkhol Osteobrama vigorsii.*
4. Large Cat fishes: *Aorichthys aor, A. seenghala, Wallago attu Pangasius pangasius, Silonia silondia, S.childrenii.*
5. Feather backs: *Notopterus notopterus, N. Chitala.*
6. Air breathing cat fishes: *Heteropmustes fossilis, Clarias batrachus.*
7. Murrals: *Channa marulias, C. stratus, C.punctatus, C. gachua.*
8. Weed fishes: *Ambasis nama, Esomus danrica, Aspidoparis morar, Amblypharyngodon mola, Puntius sophore, P. ticto, Oxygaster bacaila, Laubuca laubuca, Barillus barilia, H. bola, Osteobrama Cotio, Gadusia Chapra.*
9. Exotic fishes: *Oreochromis mossambicus, Hypopthalmichtys molitrix, Cyprinus Carpio specularis, C. Carpio Communis, Gambusia affinis, Cienopharyngodon idella.*

Source: Sugunan (1995).

Table 24A: Effect of Reservoir Formation on the Fishery of Different Streams/Rivers

1. *Himalayan streams*: The Mahaseers; snow trouts and *Labeo dero,* and *L. dyocheilus.*
2. *Ganga River System*: The anadromous Hilsa, The catadromous eels, and fresh water prawns and major carps.
3. *Mahanadi River*: *P. Sarana, T. tor. Tor mahanadicus, T. mosal, L. fimbriatus, L. Calbasa* and *Rhinomugil Corsula.*
4. *Cauvery Basis*: *P. dobsoni, P. dubius, P. Carnaticus, C. Cirrhosa* and *Labeo Kontius.*
5. *Krishna River System: P. Colus, P. dubius, P. Sarana, P. Porcellus, L. fimbriatus, L. Calbasu, L. Pangusia,* and *Tor Kudree.*
6. *North East:* The Mahseer, eels and *Osteobrama belaugiri.*

Table 24B: Effect of Impoundment on Fish Population in Parent River and Reservoir (A Sample Survey)

River	*Species of Fish Present*	*Fish Affected/Remark*
Mahandi	**Before construction of Hirakud.** Rich Fish fauna. 103 species recorded, comprising both plain and submontane forms with sizable representation of carps and Cat fishes.	Worst affected are: *T. mosal, Rhinomugil Corsula* and fresh water *Macrobrachium malcomsonii*
	In the past: Common species, *P. sarana, T. tor, T. mahanadicus, T. mosal, L. fimbriatus* and Indogangetic Carps.	
	At present: The number of species declined to 40 of which many may disappear, Mahseer.	Got stronghold after impoundment (Ravisankar sagar Reservoir)
	Cat fish (*Aorichthys aor* and *Aorichthys seenghala*) showed existence earlier.	
Cauvery river System	Mahaseer established	*P. dubius, P. carnaticus* *C. cirrosa, C. reba,* *L. Konitus, L. fimbriatus,* *T. putitora,* *Acrossochilus – hexagonolepis* are affected in various degrees by impoundments like – Krishnaraj Sagar, Mettur Dam, Bhavanisagar, and Amaravathy reservoirs.
	Hilsa illisha	Affected due to construction of anicuts.
Krishna River system: Reservoir	*Mahseer established*	
Tungabhadra Res:	*P. Sarana, Labeo spp., Indigenous, P. Kolas, P. dubius, P. Sarana, P. porcelius, P. potail Labeo pangusia,*	Most of the fish disappeared or declined drastically due to absence of fluviatile environment and changed trophic structure.
	Cat fish Established.	
	Maheseer do do	Could not come up.
	Major Corp do do	do
	Macrobrachium malcomsonii	Nor assessed.
Nagarjuna Sagar:	*Thynnichtys sand khol*	*P. sarana, Labeo spp. L. fimbriatus, L. Calbasu, T. Khudree* declined on account of recurring breeding failure. Established
Ravisankar Sagar:	Minor carp fisher was represented by *Gadusia chapra* followed by *Osteobrama cotio*	Did not appear in the river Mahanadi earlier. After impoundment provided forage base for catfish.
	Dominant Cat fish (*Aorichthys aor, Aorichthys seenghala* got strong hold after impoundment.	Catfishes existed earlier in the river.

Contd...

Table 24B–Contd...

River	Species of Fish Present	Fish Affected/Remark
Hirakud Res:		Mahseer could not establishe.
Govindgarh Res:	*Tort or* – Established.	
Bhavani Sagar Res:	Indigenous spp. like *Puntius* spp., *Tor putitora, Tor tor, A. hexagonolepis, P. dubius, P. Carnaticus, P. Kolus, C. Cirrhosa* etc. - Established.	
Jaisamand Res:	*Tor tor* – Established.	
Mettur Dam.	*Puntius* sp. 28% (1943-44)	Disappeared in 1970.
	Labeo kuntius *Cirrhina cirrhosa* Gangetic major carp. Chanos. *Etroplus suratensis.*	Could not establish.
Sardasagar Res:		Major corp could not establish.
DVC Reservoir:	Mahseer – established.	
Krishnaraj Sagar:		Mahseer could not establish.
Malampuzha Res:	*Puntius curmuca* – established.	
Mangalam Res:	Puntius Curmuca – established.	
Pawai lake:	Mahseer – established.	
Reservoirs of Kerala :	Major Corps *Peninsular carps* *Cyprinus Carpio* *Chanos chanos* *Eutroplus suratensis* *Tilapia mossambica*	All established
Matatila Res:	Cat fish (23.30%) 1963-64 – established	
Sarda Sagar Res:	Cat fish established (36.8%) 1963-64. - do - - do - (39.79%) 1964-65. - do - - do - (60.94%) 1965-66.	
Chandraprabha Res:	Cat fish established (27.20 per cent)	
Nangarh Res:	Cat fish established (27.20 per cent)	
Bhimsura Res:	Cat fish established (27.20 per cent)	
Latif shah Res:	Cat fish established (27.20 per cent)	
Nanaksagar Res:	Cat fish established (60.94 per cent) 1965–66.	

Source: Jhingran and Tripathi (1969), Sugunan (1995).

Table 25: Statewise Total Number of Reservoirs Alongwith their Corresponding Area (FRL)

State	∠1000 ha Small Reservoirs		Irrigation Tanks		1000–∠5000 Medium Reservoirs		5000–above Large Reservoirs		Total		IIMA (1983) (Additional data) 5000 ha and Above	
	No.	Area (ha)	No.	Area (ha)	No.	Area (ha)	No.	Area (ha)	No.	Area (ha)	No.	Area (ha)
Tamil Nadu	58	15663	8837	300278	9	19577	2	23222	8906	358740	44	50875
Karnataka	46	15253	4605	213404	16	29078	12	179556	4679	437291	28	151624
Andhra Pradesh	98	24178	2800	177749	32	66429	7	190151	2937	458507	117	198330
Gujarat	115	40099	561	44025	28	57748	7	144358	711	286230	22	125119
Madhya Pradesh	6	172575	–	–	21	169502	5	118307	32	460384	56	250000
Uttar Pradesh	40	20845	–	197806	29	44993	4	71196	66	334840	98	445000
Maharashtra	–	119515	–	–	–	39181	–	115054	–	273750	72	105202
Bihar	112	12461	–	–	5	12523	8	71711	125	96695	23	50000
Orissa	1433	66047	–	–	6	12748	3	119403	1442	198198	15	99061
Kerala	21	7975	–	–	8	15500	1	6160	30	29635	–	–
Rajasthan	309	54231	–	–	30	49827	4	49396	423	153444	32	120000
Himachal Pradesh	1	200	–	–	–	–	2	41364	3	41564	3	40000
West Bengal	4	732	–	–	1	4600	1	10400	6	15732	7	15735
Haryana	4	282	–	–	–	–	–	–	4	282	–	–
North East	4	1639	–	600	2	5835	–	–	6	8074	–	–
Total	2331	551695	16803	933862	180	527541	56	1140268	19370	3153366	517	1650946

Source: Sugunan (1995) and IIMA (1983).

Diverse opinion persists about the classification of Indian reservoirs. Fish seed committee of Government of India (1966) termed water bodies of more than 200 ha in area as reservoirs. (David *et al.*, 1974) while classifying reservoirs of Karnataka state considered impoundments above 500 ha as reservoirs and named smaller ones as irrigation tanks. IIMA (1983) on the other hand classified reservoirs of India into 3 categories as (1) impoundment having more than 5000 ha area are considered as large, between 1000–5000 ha area as medium and below 1000 ha area as small reservoirs. But the problem remains about the small reservoirs. In states located in Peninsular India like Andhra Pradesh, Karnataka and Tamil Nadu, small reservoirs are also known as irrigation tanks and most of them are seasonal but contributed fishery applicably. Hence, two avoid all the anomalies, the irrigation reservoirs or tanks are shown separately in the same Table 25 in this document and reservoirs from 10 ha to ∠ 1000 ha are considered as smaller reservoirs as shown by Sugnnan (1995) 1000–∠ 5000 ha as medium and 5000 ha and above as large reservoirs. Statewise total number of reservoirs of different categories together with approximate areas (ha) and of irrigation tanks are shown in table 25. Additional information on larger reservoirs > 5000 ha area as indicated by IIMA (1983) are all also shown in a separate column in the same Table 25.

In order to obtain some idea about the production characteristics fish, random sampling of some reservoirs irrespective of their zones are grouped into five categories. These are 10–∠ 500-ha; 500–∠1000 ha; 1000–∠10,000 ha; 10,000–∠ 20,000 ha and 20,000–above along with their corresponding production value (average) with ranges from the total number of reservoirs surveyed are indicated below in Table 26.

Table 26: Production Characteristics in Different in Reservoirs of Different Size Groups Ranging from 10-70,000 ha

Reservoir Size (ha) at FRL	*Average Production t/yr.*	*Range*	*Number of Reservoirs Surveyed*
10–∠500	18.22	0.1-380	130
500–∠1000	24.406	0.25-450	32
1000–∠10,000	43.317	0.3-580	8
10,000-∠20,000	320.35	40.5-964	4
20,000–Above	773.528	14.56-3085	5

Source: Based on the date of Sugunan (1995).

It is interesting to note that although the upper level of production indicates fairly high value in all the categories of reservoirs, their corresponding lower level is alarmingly low. The low values are recorded in a number of reservoirs. This is evident from all the 5 groups considered here. Out of 130 reservoirs (10-∠500 ha area) when the production ranged between 0.1-380t/yr. The average value shows only 18.22 t/yr. Likewise in case of 500–∠1000 ha area, the average production is increased to 24,406 t which ranged between 0.25-450 t in case of 32 reservoirs. In reservoirs 1000–∠10,000 ha area, 10,000–∠20,000 ha area and 20,000–above where production ranges were 0.3-580 t; 40.5-964 t and 14.56-3085 t from 8, 4 and 5 number of reservoirs, their corresponding average production values are only 43.317t; 320.35 t and 773.528 t respectively.

Statewise distribution of total number of reservoirs small (∠10000 ha area), medium (1000–∠5000 ha area) and large (5000 above) with their total production are given in Table 25. This will help in finding out the national production rate of fish in kg/ha.

Sugunan (1995) has shown fish production particulars of 422 reservoirs (291 small, 110 medium and 21 large reservoirs) from only 10 states (Table 27) during 1992-93. This has enabled to bring out an approximate national fish production rate from the reservoirs. Additional information on fish production only in case of large reservoirs from the states Gujarat, West Bengal, Haryana and North east ranges which are also available are not incorporated here, only their numbers and the corresponding sizes are shown in the Table 25. Table 27 indicates that the average fish yield of small, medium and large reservoirs are respectively 49.90, 12.30 and 11.43 kg/ha. The National fish production rate calculated from the table works out to be 18.12 kg/ha (not 20.13 kg/ha) as reported by Sugunan, 1995). By applying this production rate (kg/ha) obtained from 10 states (leaving values of Gujarat, West Bengal, Haryana and North East Zones) into the total area (ha) of these groups of reservoirs *i.e.* 551695 ha, 527541 ha and 110268 ha respectively (Table 25) for small, medium and large reservoirs, the production figure works out to be 27529 t, 6488 t and 130335 t respectively. In case irrigation tanks below ∠1000 ha shown in Table 25) are also included, the production figure comes to 46599t. The inclusion of production figure from the additional 4 states as stated would have given more correct value of National Fish production rate in place of 18.12 kg/ha as shows in Table 27.

Further IIMA (1983) have stratified reservoir resources into three main categories. These are larger–more than 5000 ha in area; medium between 1000–∠5000 ha and small–below 1000 ha area. According to the latest report (IIMA, 1983) there are 975 reservoirs in the country. Out of this there are 550 major reservoirs. Of this 5 per cent are very large (above 10,000 ha), 3 per cent large (5000–∠10,000 ha). This means in all 5 per cent + 3 per cent = 8 per cent are above 5000 ha; 26 per cent medium (1000–∠5000 ha) and the rest are small. From Table 25 shown by Sugunan (1995) it is clear that the number of smaller and medium reservoirs (2331 + 180) are considerably high. But in case of larger reservoirs the number shown is much less (only 56) than what is shown by IIMA (1983), which is 550. This means the increased production value of extra (550-56) = 494 major reservoirs are not incorporated which otherwise would have enhanced the fish yield (kg/ha) value. Sugunan (1995) has shown the total areas of larger reservoirs (>5000 ha) as 1140268 ha from 56 reservoirs (Table 25) against 1650946 ha from 517 reservoirs (IIMA, 1983) also shown in Table 25). Calculating this value of IIMA (1983) based on the average value 11.43 kg/ha (Table 27) the total production value (shown by Sugunan (1995) is 13033t against the value now calculated is 188705. This means an additional increase of 5837t from the larger reservoir resources are left out. Thus the yield, area and the production (Table 28) comes out as shown in (Table 28) additional information on fish yield in some reservoirs are also shown in Table 29.

Table 27: Statewise Total Number of Different Groups of Reservoirs with their Corresponding Fish Yield

State	Smal Reservoir (∠ 1000 ha)			Medium Reservoir (1000–∠ 5000 ha)			Large Reservoir 5000 and above			Pooled			Large Reservoirs 5000 and Above (ha) Additional Data IIMA(1983)		
	No.	Production (t)	Yield kg/ha	No.	Production (t)	Yield kg/ha	No.	Production (t)	Yield kg/ha	No.	Production (t)	Yield kg/ha	No.	Production (t)	Yield kg/ha
Tamil Nadu	52	760	48.50	8	269.0	13.74	2	294.0	12.66	62	1323	22.63	44	1552	30.6
Uttar Pradesh	31	168	14.60	13	156.0	7.17	1	50.0	1.07	45	374	4.68	98	1996	4.5
Andhra Pradesh	37	2224	188.00	29	306.0	22.0	3	800.0	18.80	69	333	36.48	117	4386	36.9*
Maharashtra	6	72	21.09	12	313.5	11.83	4	794.0	9.28	22	1179.6	10.21	–	NA	NA
Rajasthan	78	970	46.43	17	599.7	24.47	2	120.0	5.30	97	1690	24.84	–	NA	NA
Kerala	7	118	53.50	2	17.3	4.80	–	–	–	9	135	23.37	–	–	–
Bihar	25	22	3.91	3	7.2	1.90	1	0.8	0.11	28	30	0.054	–	NA	NA
Madhya Pradesh	2	24	47.28	20	624.9	12.02	3	1184.0	14.53	25	1833	13.68	56	2164	8.6
Himachal Pradesh	–	–	–	–	–	–	2	1453.0	35.55	2	1453	35.55	3	1300	32.5
Orissa	53	349	25.85	6	163.0	12.76	3	925.0	7.62	62	1437	9.72	15	950	9.6
Total	291			110			21								
Average			49.90			12.30			11.43			18.12			

Source: Sugunan (1995).

*: This includes production of only 97 reservoirs comprising of 118728 ha. The productivity for Andhra has been calculated using the total area of these 97 reservoirs.

Table 28: Yield, Area and Production from Different Categories of Reservoirs

Category of Reservoirs	*Area (ha)*	*Yield kg/ha*	*Production (t)*
Small Reservoirs	551695	49.90	27529
Irrigation Tanks	933862	49.90	46599
Medium	527541	12.30	6488
Large	1140268	11.43	13033
Additional increase from large reservoirs	510678		5837
Total	3664044		99486

Table 29: Additional Information on Fish Yield in Some Reservoirs

State	*Reservoir*	*Area in (ha)*	*Fish Yield kg/ha*	*Period*	*Source*
Haryana	Bhadkhal	761	8	1979–80	IIMA (1988)
Uttar Pradesh	Sarada Sagar	4885	32	1980–81	–Do–
	Pili	1184	16	1979–80 1980–81	–Do–
	Dhora	1280	47	–Do–	–Do–
	Rihand	46539	10.91	–Do–	Natarajan (1979) in Jhingran (1991)
Madhya Pradesh	Maniari	2526	33	1979–80	IIMA – (1988)
	Dohart	604	7	1979–80 1980–81	–Do–
Himachal Pradesh	Govind Sagar	16867	76.5		Jhingran (1991)
Maharashtra	Joyakwadi	39777	9	1979–80 1980–81	IIMA (1988)
	Itiado	7410	10	–Do–	–Do–
	Panset	1252	11	–Do–	–Do–
Rajasthan	Jaisamond	7286	116	1979–80	Dubey (MS)
	Ramgarh	1267	104	–Do–	–Do–
Andhra Pradesh	Kelleru	69600	197	1979–80 1980–81	–Do–
	Nizam Sagar	28475	10		Natarajan (1979) in Jhingran (1991)
Gujarat	Kodana	18636	6	1980–81	IIMA (1988)
	Panam	18740	6	–Do –	–Do–
	Bhadar	5960	8	1981	–Do–
Bihar	Bhadua	1336	8	1979–80 1980–81	IIMA (1988)
Karnataka	Tungabhadra	37814	57	1979–80	Dubey (MS)
	Kabini	6092	70	–Do–	–Do–
Orissa	Hirakud	71400	11	1979–80 1980–81	IIMA (1988)
	Bhanjangarh	600	104	–Do–	–Do–
Tamil Nadu	Santanur	2010	88	–Do–	–Do–
	Bhavani Sagar	7876	80	–Do–	Natarajan (1979)

Source: Dubey, G. P. (MS) Mameo.

Again since irrigation reservoirs have fluctuating water levels and therefore according to IIMA (1983) the area for the purpose of production per ha has been as 60 per cent of PRL (Full Reservoir level), NCA (National Commission on Agriculture) (1976) has suggested that for uniformity the average area between FRL (Full Reservoir level) and DSL (Dead storage level) should be considered for calculating the yield/ ha.

According to NCA (1976) the average fish production from Indian Reservoirs was frustratingly low *i.e.* 5-8 kg/ha. IIMA (1983) later made a sample survey of 32 reservoirs of different categories and found that it varied from 1-120 kg/ha. The average production however reached 14.5 kg. According to statistics collected by IIMA (1983), reservoirs in Andhra Pradesh showed the overall production of 36.9 kg/ha followed by Tamil Nadu 30.6 kg; Rajasthan (as stated by the Government) was about 30.0 kg/ha (much higher than table 26). Other states showed lower productivity, the lowest being West Bengal.

Though the average production in India is low, but, some of the reservoirs have now attained very good production. For example, "Govind Sagar (11,122 ha) has yielded 70.30 kg/ha; Tungabhadra (22,688.4 ha) and Kabina (3,700 ha) in Karnataka have yielded 95,57 and 114,5 kg./ha respectively. Jaisamand (4,370 ha) in Rajasthan has yielded as high as 187.85 kg/ha.

Further production of fish indicated in Table 29 in some reservoirs gives additional information available. Such values are not clearly projected by Sughan (1995) in his document, such additional information on production may help to project higher values in the National Yield (kg/ha) in addition to the values indicated in Table 26.

Chapter 16

Escapement of Fish from Waste Weirs and Canals

Waste weirs and sluice gates of Canals in reservoirs are the sources through which considerable loss of fishery takes place during monsoon when the reservoir water touches the highest level (Dubey Ms). This is also one of the reasons of low production of the reservoir. An attempt was made to study the loss of fish from a reservoir in Pelwa dam in Madhya Pradesh during 1975-76. It was reported that about 7.30 tonnes of fish escaped and this works out to be 36 kg/day. The author further states that fish escapement through canals is also considerable. However, a detailed study on this aspect is essential in various reservoirs along with other studies. In case such escapement is appreciable, suitable screens or other suitable measures are to be thought of to prevent escapement. Dubey (MS) reports that installing trush racks like these fixed in the head of the main canal in Tawa and general reservoirs in Madhya Pradesh has helped in checking fish escape. It is reported that electric barriers seem to be better alternative for this purpose. USA, FRG, Russia, Poland and other countries have done considerable progress on this direction.

Chapter 17
Poaching

The extent of poaching affects the net production of fish from reservoirs. An estimate reveals the extent of annual poaching in Rihand reservoir (46538 ha) which is about 65 t/year, 26 t in Kangsabati (10,400 ha), 15 t each in Govind Sagar (16867 ha) and Hirakud (71963 ha) reservoir. In fact poaching has been increasing in most of the reservoirs. In smaller reservoirs poaching is rather easy and the entire production can be removed in a few nights. Hence to stop such practice stern action has to be taken by imposing heavy penalty.

Chapter 18

Breeding Grounds

Natural breeding grounds in the reservoir, their confluence with the river or in the mainstream from where seed resources are brought to the main reservoir is a natural phenomenon. Such natural stocking has considerable importance in maintaining the productivity of reservoirs provided the brooders are of commercial importance and can compete with the undesirable non-commercial croup of fishes. Since most of the reservoirs have come up around 70s and onwards, not much of work could be planned on this aspect.

Although it is known that many of the major carps will be able to adopt themselves to different spawning conditions and may under certain circumstances breed in lakes and reservoir (Hora, 1945), one aspect of the problem which has not drawn serious attention of workers in the breeding of major carps in reservoirs. It would be interesting to find out in each reservoir suitable for fishery development, whether in case of fertilized eggs or spawn produced in it do not survive and thus result in natural stocking, the particular location can be used as fish seed collection centre.

Although it is an established fact that dynamics of spawning and Prerecruitment of carps is governed by variety of complex factors like meteorological, environmental and biological, but the fact is that it requires continued studies for few years to elucidate the influences and interrelationships of these factors for successful spawning success between reservoirs.

Ramchandra Menon (1969) recorded that, of the eighteen hydel reservoir in Tamil Nadu state, it has not been possible to pin point the breeding grounds of major carps stocked in them. In some, there is natural replenishment of stocks while in others planting of fingerlings year after year have to be resorted to. In this opinion major carps do not spawn in deep trough like lake basin of reservoirs which is their trophic environment but migrate to shallot upstream riverine area.

It is thus more or less established that mature major carp breeders normally do not breed in the main reservoir and prefer migrating up-stream for the purpose of breeding.

Further, exposure and destruction of the desmersal eggs arising from frequent marginal exposure of the river bed at–short interval of time, eggs getting washed into deeper portion of the reservoir where they get destroyed and the loss in the maturity condition of fishes, as breeding is rendered impossible in the frequently exposed habitat, caused by extreme and abrupt flood fluctuation (characteristic of DVC reservoir). Sudden lowering of temperature due to heavy rainfall seems to affect the maturity condition of fish. Malhotra *et al.* (1966); Govind and Khan (1969) had similar observations that abrupt fall in water level alone, yields, eggs which are drawn down into the river course in the receiding floods before they could become hatchlings. A spawning ground of major carps has been located at the confluence of Bhadra river with the reservoir in Bihar (Ghosh and Rao, 1969). The collection of fingerlings of mrigal and Catla from Mettur Reservoir and the age-group wise of fishery, indicate that these species have established themselves well in the reservoir.

Some observations of fish breeding recorded by various workers in different reservoirs are indicated below:

Investigation carried out by Gopalkrishnan *et al.* (1966) have shown that Catla, rohu and mrigal spawn in the tail ends of Tilaiya reservoir while mrigal and Calbasu in Panchet Reservoir.

Large scale escapement of matured fishes from the Malampuzha reservoir into the river is noticed during South-West monsoon periods (June and September) when the spill-way gates of the dam are open to regulate the storage of water in the reservoir (Nair, 1969).

Ramchandra Menon (1969) observed that in case of Bhavanisagar reservoir mass occurrence of fertilized eggs of major carps of Bhavani river about 6 miles up stream brought into the reservoir proves that successful repopulation of carps depend on the accessibility to a spawning run in swelled river in the top waters of the reservoir in the low flood characterised by high turbidity and when the migratary facility is absent, carps do not propagate. The brood stock ascends about 35 miles up stream from the reservoir water when the fish flood mingle in the reservoir water mass spawning occurs in shallow sandy river margin.

The fact that major–corps mostly *C. mrigala* and *C. Catla* breed at the head water region (near Parbara) is also recorded by Gopalakrishnan and Pal (1969). The actual spawning activities during their investigation carried out in Tilaiya Reservoir could be observed and about 104 lakh carp eggs were collected in 3 days in Midnapore type collection nets.

In the Panchet reservoir spawning of carps takes place at headwater region and in the neighbouring areas of Damodar river, but the quantity of spawn that could be collected was not much which evidently was due to large scale destruction of breeders by local people. Species composition of spawn collected was *C. mrigala*, 83.2 per cent, *L. bata* 10.1 per cent; *L. rohita* 4.0 per cent and *C. reba* 2.7 per cent. Spawning ground of major corps has been located at the confluence of Bhadra river with the reservoir in Bihar (Ghosh and Rao, 1969).

It is thus evident that from the above findings of various workers that breeding occurs in general in the shallow areas of head water river or in the confluence of river and the reservoir. But in the shallow zone of the main reservoir, breeding cannot be rutted out. The desired quantity of spawn/fingerlings which should have shown considerable stocking intensity is not possible due to the following factors:

1. Accessibility of mature breeders while migrating from reservoir to head waters are largely captured and destroyed (Mahanadi and Ib river of Orissa).
2. Abrupt and full in water level alone yiels mass occurrence of fertilized eggs which are brought into the deep waters during the receiving floods of reservoir where they can not hatch and are destroyed.
3. The loss in the maturity condition of fishes, as breeding is rendered impossible, in the frequently exposed habitat caused by extreme and abrupt flood fluctuation.
4. Sudden lowering of temperature due to heavy rainfall may affect the maturity condition of fishes (Malhotra *et al.*, 1966).

Chapter 19

Stocking of Reservoirs

From the foregoing account due to uncertain stocking from natural breeding it is thus evident that the stocking of fish seed (especially of commercial important) in the reservoirs has become essential to obtain higher production. This is an age old practice in the country for stocking ponds. It is from the time the importance of fish culture is considered as essential, the stocking practice was initiated. But the practice of stocking of fingerlings of a single species or a combination of species (with ratio) without any scientific base have resulted in unsatisfactory production. The seeds of indigenous species collected from nearby rivers were directly stocked in the reservoir. Such mode of stocking used to contain commercial important species along with larger percentage of undesirable weed fishes and predators. As a result the weed fishes used to compete for food. likewise predators feed on small fry of commercial species. Further, stocking rate was also without any base. Such non-scientific stocking eventually did not enhance fish productions.

Again with the advent of "induced breeding" most of the states in India are to raise the carp seed in large numbers and this has resulted in to produce enough of Gangetic major corps like Rohu, Catla and Mrigal, the most commercial important species for stocking in various reservoirs (both in Gangetic and peninsular reservoirs). But simultaneously it is worth while, to give priorities to the liking of the local varieties.

Impact of Stocking

The stocking of reservoirs would have been a successful and meaningful if the stocked fish could have attained a healthy growth, mature and properly reproduce. In many cases it has been observed that despite persistant stocking, the transplanted species did not show the desired result (proper growth and reproduction). Smaller the reservoir better are the stocking and recapture. The following are some of the results obtained by some scientists (Sugunan, 1995).

Table 30: Showing Effect of Stocking in Reservoirs

Sl.No.	Reservoir	Species Stocked	Results
1.	Mettur (19346 ha) (Stanley R) (Tamil Nadu)	*C. Catla*	Naturalised population periodic set back on two occasions due to breeding failure.
		L. rohita	Adequate.
		L. Calbasu	
		L. fimbriatus (Common carp) *L. Kontius* *P. Carnaticus* *C. reba* *P. dobius*	Never recaptured in appreciable quantity.
2.	Bhavani Sagar (7876 ha) (Tamil Nadu)	*C. Catla*	No impact on catch structure
		L. Calbasu *L. fimbriatus*	Established and found the environment congenial for breeding. Did not make any place in fisheries although its was native to the ecosystem.
3.	Santanur (1255 ha) (Tamil Nadu)	*C. Catla*	Breeding has been successfully established. Contributed 80% of total catch.
		L. fimbriatus *C. Cerrhosa* *L. Kontius* *C. reba*	Effectively reduced due to inability of fishes to breed and propagate.
4.	Vaigal (2420 ha) (Tamil Nadu)	Gangetic Carp *L. fimbriatus*	Stocking is a waste as they could not contribute to the total fishery.
5.	Krishnagiri (1250 ha) (Tamil Nadu)	Indian major Carps	Could not come in the total fishery
6.	Malampujha (2313 ha) (Kerala)	Gangetic major carps	Could not show substancial increase in yield rate due to breeding failure.
7.	Peechi (1515 ha) (Kerala)		
8.	Nagarjuna Sagar (28475 ha) (Andhra Pradesh)	Regular annual stocking comprising Indian major corpps	Adsence of breeding could not contribute to recruitment.
9.	Tungabhadra (37814 ha) (Karnataka)	–	Stocked fished failed to breed.
10.	Krishnaraj Sagar (13200 ha) (Karnataka)	–	–Do–
11.	Gandhi Sagar (66000 ha) (Madhya Pradesh)	–	Stocked fishes contributed to the fishery through breeding and recruitment, Fish yield rose from 0.51 kg/ha to 20.33 kg/ha. Mrigal despite being indigenous ichthyo fauna declined.

Contd...

Table 30–Contd...

Sl.No.	Reservoir	Species Stocked	Results
12	Ravisankar Sagar (9540 ha) (Madhya Pradesh)	Carps	Could not establish due to breeding failure of stocked fish.
13	Ukai (36523 ha) (Gujarat)	Indian major carps	Breeding success resulted as raise of production. But young fishes are reported to escape through outlet of the Dam.
14	DVC Reservoirs Tilaiya (6475 ha) Maithon (14491 ha) Panchet (7511 ha) (Bihar)	-	Stocking did not have any impact.

From the above observations of stocking in reservoirs irrespective of their sizes and location it is clear that mere stocking of reservoirs with spawn/fry does not always mean that the spawn/fry will survive, have a healthy growth, attain maturity and have a natural breeding, nevertheless in some of the reservoirs breeding success is established.

In case of smaller reservoirs stocking can be made more effective through proper management and harvesting and thus higher production is resulted. This has nothing to do with their breeding success/recruitment. Hence. stocking stould be done in all types of reservoirs adequately and scientifically to get adequate yield.

This is why possibly IIMA (1983) during their study have point out that the reservoirs should be systematically changed from capture to culture fisheries by stocking them with appropriate number and size of fingerlings of fast growing carps, as they utilize different ecological nitches of reservoirs of all categories, Smaller reservoirs can yield two to three times more *i.e.* 2 50 to 500 kg/ha and the medium reservoirs from 150 to 2 50 kg/ha. As such it would be possible to increase the present production from 25000 t (VII Flan GOI, 1984, 14.5 kg/ha) to 1,20,000 t (40 kg/ha) in total water area of 3 m ha.

This shows that low production indicated in smaller reservoir (Table 27, stated earlier) were not adequately stocked and scientifically managed in the past which resulted alarmingly low production. This is why capture-cum-stocking basis has gained importance. Such practice will gradually encourage the culture fishery practice in reservoirs in place of capture fishery.

Diverse opinion exists with regard to stocking of fingerlings in reservoirs of different categories. Hey (1952) suggested 100 fingerlings/ha in larger reservoirs. Huet (1960) and Jhingran (1969) have suggested certain formula for stocking reservoirs. IIMA (1983) has suggested stocking of fingerlings (12.5 cm size) @ 200/ha for Larger reservoirs (above 500 ha), with 500 fingerlings for reservoirs between 1000 to 5000 ha and 1000 fingerlings/ha for reservoirs less than 1000 and 2000 fingerlings between 25-50 ha area. FSC (1966) has suggested 495 fingerlings/ha of the size of 40-150 mm size. NCA (1976) suggested 500 fingerlings/ha, whereas Dr. Dwivedi/suggested 500-1000 fingerlings/ha. This shows that requirement of fingerlings is considerably high which needs adequate effort to produce and stock. It is suggested that keeping the above suggestion in view the number of fingerlings that should be stocked in different. Category of reservoirs are.

1. Large very large reservoir (5000–above) stocking should be 200–500 fingerlings/ha.
2. Medium reservoirs (1000–5000 ha), 500–1000 fingerlings/ha.
3. Small reservoirs (∠1000 ha), 1000–2000 fingerlings/ha.

According to report of IIMA (1983) there are approximately 555 fish seed farms which are mostly rearing farms, but the percentage contribution for reservoir stocking is hardly 13.5 per cent where as that of ponds and tanks are 89.7 per cent.

Further, it is difficult to have proper size and adequate number of fingerlings/ha for stocking. Hence, a set of fingerlings ponds are to be set up in the vicinity of reservoirs to facilitate raising of proper sized fingerlings.

But mere stocking straight way does not help unless some detailed information about the existing reservoir fishery is known and based on that only stocking may be made properly. Such aspects are highlighted during the study of four DVC reservoirs by Jhingran and Natarajan (1969).

Tilaiya Reservoir

Completed in 1952. If the fishery of Tilaiya depends entirely on the stocked population (Main Catla) there should exist a direct relation between fishing intensity, stocking density and yield. In the year 1964 the catch per unit of effort in Tilaiya reservoir was 0.640 kg but in 1958 the catch by the same party was only 0.146 kg. There are many identical aspects in the fishery between these two periods. Catla formed 88 per cent of the catch in 1964 and 89 per cent in 1968, the average weight of fish being 5.36 kg in 1964 and 4.18 kg in 1968. The yield was found to be 24 m tons in 1964 but the same fell down to 1/12th in 1968 for about a third of 1964 effort. The data thus clearly indicate that the population of Catla got considerably thinned from 1964 stock density and required replenishment. This also incidentally indicates of breeding. Again examining the fish population in general the catch was 32.6 m tons in 1963-64. From an effort 50 per cent more in 1967-68 the yield fell by about 1/3. This may mean that there is a trend of reduction in population densities of other fishes as well. In the absence of non-availability of qualitative composition of stocked fingerlings, comments on self-stocking capacity of fishes other than Catla is not warranted. In the year 1960, the species composition, in order of abundance was Mrigal, Rohu, Catla and Calbasu. In 1967–68 Catla and Rohu interchanged position indicating changes in population structure. So while stocking, all such points should be taken into consideration.

Maithon

Completed in 1957. Stocking with major corp fingerlings commenced in the year 1958 and continued in successive years unabated. The recruitment on the basis of size frequency was maintained by stocking as it marked the recruitment and thus the breeding. Only negligible percent of mrigal available in the rared egg samples was the only available information on breeding. If that is so there should be some relation between stocking density, fishing intensity and catch. The catch per unit of effort in 1964 for Maithan was 0.234 kg while in 1968 it was 0.194 kg. In both the periods Catla was dominant forming 97 per cent and 99 per cent of the catches respectively. Average size of Catla in the Catch was also more or less same, 8.7 and 7.07 kg respectively. Scale reading as per sample collected in 1968 indicates that fishes of those average weights may be between 5 and 6 years old. The enhanced stocking in 1961 to 1963 thus has not been felt in Catches in 1968. Examining the fishing effort the catch/unit of effort in 1962-63 was 0.322 kg. While in 1967-68 it was 0.195 kg. In 1967-68 the effort was about 150 per cent more than 1962-63 but the catch increased only from 9.1 m tons to 13.43 m. tons. All these observations clearly indicate that stocking has had little effects on production in Maithon reservoir. It is therefore felt that it would be proper to suspend temporarily stocking Maithon reservoir for a couple of years which is the only way to know whether there is recruitment or otherwise. Unless this point is fully clarified there is little justification in continuing stocking of Maithon reservoir. Since Maithon reservoir has been stocked in successive years enough to give much better production than what is yielding presently; a two year reprieve in stocking may not affect the fishery too adversely but would provide scientific information for proper development of the fishery. The species composition for 1966-67 and 1967-68 indicated that Catla formed 64 to 67 per cent by weight, mrigal 20 to 22 per cent, rohu 2 to 6 per cent Calbasu 1 per cent and common carp 0.2 per cent. It is worth examining whether it is desirable to stock the common carp which form such meager catch. Here also whether natural breeding occurs or not is an essential information. All these information should have some relation between stocking density, fishing intensity and catch as explained above.

Panchet

Completed in 1958. Stocked only in 1958. The same recurring dominant sizes in different years namely 15"-16" (Catla), 17"-18" (rohu), 20"-22" (mrigal) and 16"-18" (Calbasu) in 1960 and 1961 suggest recruitment of these fishes. The occurrence of these species in all years leave little doubt about their recruitment and breeding. Available data and observations point to breeding and recruitment of all major carps. The species composition of fishery in early years (1960 and 1961) indicates, among major carps a dominant position of mrigal followed by Rohu. In 1967-68 mrigal remained the dominant species but Catla took the second position followed by Rohu. In overall species composition including cat fishes *M. aor* takes the first place from 1966 while it occupied the third place in 1960. This incidates a faunal change in favour of catfishes. An examination of catch and catch per unit of effort during the period 1962-68 indicates a gradual increase in the density of fish populations until 1966. Thus for comparatively largest fishing effort in 1963-64 the yield was 21.6 m tons with a catch per unit of effort as 0.196 kg. For nearly half of that fishing effort in 1965-66 the yield mounted to 38.1 m tons with catch per unit of efforts rising to 0.605 kg. In the year 1967-68 with more effort than 1965-66, there was a steep fall in production at 10.5m tons with a catch of 0.121 kg per unit of effort.

This sudden drop in production appears to be due to the fact that the reservoir was drained to ½ square mile (maximum area being 60 sq.m.) in 1966 resulting in the mass escape of fishes from undersluices and further diminution in the size of the populations from fishing mortality in that year. This has also reduced the average size of fishes like Catla to less than 2 kg in 1967-68. It is necessary that Panchet reservoir should be stocked for a couple of years and capture of major carps less than 2 kg should be stopped until the stock re-establishes themselves.

Konar

Completed in 1955. Stocking commenced in 1957 and continued upto 1959 with Gangetic major carps 5.5 lakh fingerlings. In 1962 and 1963 about 1 lakh and 9 thousand fingerlings of common corp were also stocked. During 1962-63 the catch per net was 0.353 kg when the total catch was 5.6 m tons. In 1963-64 and 1964-65 with the increased fishing effort the catch went upto 23 m tons in each of the above years and catch per unit of effort also was higher than that in 1962-63. However, in 1967-68 with nearly identical fishing effort as in 1964-65 the production fell to 10 m tons and catch per unit effort to 0.226 kg net. This could arise only from reduced stock. Catla was the most dominant species in the catches in the year 1964 with catch per unit of effort being 0.840 kg per net with the average weight of the species caught being 6.09 kg. Catla was again the dominant species in the catches in 1968 but the catch per net dwindled to as low as 0.075 kg and so did the yield (1m ton). The important feature of the catch (including Mettur fisherman party) in 1968 was that the average weight of catla went up to 12.66 kg with a modal value of length at 900 mm. and according to tentative estimates (from scale reading) the average age of Catla in the catch may be 7 to 8 years old. All the above observations indicate that the stock of Catla in the reservoir is thinning out with average size and age of the catch mountings. An examination of length frequencies lends weight to the above conclusion. There is a marked increase in the overall size of Catla from year to year which was 29" in 1963, 30" in 1964, 32" in 1965, 32 to 33" in 1966 and 36" in 1968. The above observations however do not totally preclude breeding of Catla in Konar reservoir. In fact there is a sign of feeble recruitment in 1964 as is seen from the size frequency studies. However occurrence of the same modal values in gill net catches from year to year noted in case of mrigal, rohu and Calbasu but not in Catla indicates recruitment in the former. A dominant size occurs in 1963, 1964 and 1965 in the case of rohu, 20" in 1963, 1964, 1965 and 1966 in case of mrigal and 16" in 1963, 1964 and 1965 in case of Calbasu.

An examination of species composition of catches from the year 1963 also gives clue as to recruitment and breeding. Thus in 1963-64 mrigal formed just 1.30 per cent of the fishery. Since there was no stocking from 1960 onwards and its average weight in 1967-68 has been less than 1 kg there is little doubt about their breeding and recruitment. The reverse however was true in case of Catla. This fish was dominant in 1963-64 perhaps because of earlier stocking. It is since dwindling in catch with average size of the fish mounting upwards year by year showing little, recruitment. This is how by critical examination of the existing stock individual growths in different species together with their recruitment can be known and accordingly the stocking can be done appropriately.

The above four reservoirs have highlighted their individual problems which may arise in case of other reservoirs spread over the country. Their individual problems are to be analysed prior to their stocking.

Chapter 20

Prospects of Pen and Cage Culture

To have effective stocking in reservoir establishment of fish farm/hatchery is essential.

Fish farm contains few large tanks and a number of cemented cisterns of different sizes. Broad fishes are stocked in large tanks. Eggs and hatchlings obtained from breeders are reared in cemented tanks up to a suitable size before finally stocked in the reservoir. While the hatchery designed by World Bank Project. Will have the seed farm having one circular breeding pond, 2-4 hatching ponds, number of brood stock ponds and cemented ponds for raising the spawn to fry and earthen mursery and rearing ponds to raise generally standard fingerlings of 50 mm size with all arrangements of overhead tanks, water supply etc. In pond culture these are the common practices for raising eggs and hatchlings to the fingerlings stages.

As stated earlier that scientific investigation on reservoirs was initiated only from seventy onwards, augmentation of fish seed production for stocking the reservoirs is thought of very recently. In case of shallow lakes produced by receiding floods of the river, Cage and pen culture operations are also practice to increase the seed resources to a certain size for stocking in lakes or reservoirs. Cage culture in still water (especially large fish culture ponds) are common in recent years and considerable work is done by various workers with different species of fishes Govind (1983.Ms) with *Cyprinus Carpio*: Govind *et al.* (1988); Kumaraiah *et al.* (1991); Bandopadhaya *et al.* (1991); Kumaraiah *et al.* (1986); Kumaraiah and Parmeswaran (1985); Murugesan and Kumaraiah (1978) etc.

Pen culture in ox-bow lakes are not only most profitable but can also be used as nurseries where seeds of small fish of commercial important species can be reared upto immature juvenile stage safely in a short period under a highly fertilized ecological condition (Ray, 1997). Thus pen and Cage experiments can be tried for a shot duration in large shallow bays of reservoirs where highly fertilized ecological conditions exists, keeping in view that safe water level persists for a short duration inspite of its draw down. Such period may be between September/October-January/ February (Depending upon rise of water level which is dependant upon rainfall). Hatchings attain upto small fry and fingerling size can be transferred in pens to grow upto 50 mm. size for final stocking in the reservoirs.

Pen culture practices are common in still water and also in estuary where receded water again pushed along with high tide. But in reservoirs water after attaining its highest level remains almost stable for a short period after which shows gradual fall. Hence the period September/October–January/February (3/4 months) is suggested for stocking when wind velocity decreases considerably.

Singit (MS) 1983 conducted fish seed rearing in pens constructed in Tungabhadra Reservoir during 1982-83 and 1983-84. During the 1st year the overall percentage of survival was only 10.60 per cent. The technology adopted during the 1st year was modified to get a better percentage of survival of about 28.3 per cent.

A single inner pen with a dimension of 70 x 70 cm having an area 4900 sq.m. covered by 30 mesh monofilament fabric is partitioned with 25 compartments of about 200 sq.m. each with date palm (toddy) leaf mats.

One 30 mesh monofilament fabric was used for covering the inner pen. Stocking of seeds made in different chambers, but later it was found that large number of predatory weed and other unwanted fish and insects were noticed in the inner pen.

Submerged aquatic weeds such as chara, Ceratophglun, Najas and algal blooms were noticed in abundance. This may be due to use of organic and inorganic manure, poultry manure, fish feed etc. used for the spawn/fry in excess.

The system was however modified. The outer pen was extended by about 25 m to 325 metre bringing more are from 0.49 ha to 0.8625 ha. The entire outer pen is protected by covering with 30 mesh monofilament fabric and out of the independent inner pen 4 chambers were covered with 40 mesh monofilament fabric and 3 chambers with 30 mesh monofilament fabric. While erecting the outer and inner pens, a trench of 2 ft wide and 1 ft deep was made and the soil shifted at the outer edge and then pits of 2 feet deep were made along the inner edge of the trench at a spacing of 5 ft cuts in which casorina poles were fixed. Bamboos were tied to casurina poles horizontally from the ground level at a spacing of 2 ft to 3 ft to an average height of $5^1/_2$ ft and then the mono-filament fabric which was already stiched to the uppermost required bamboo line and the fabric is spread down along the Casurina and bamboo frame towards the ground and into the trench and the remaining of the fabric is spread in the trench held tightly and the dugout earth is filled into the trench upto ground level so that at least 1 to $1^1/_2$ ft of the bottom and of the fabric is securely embaded in the soil so that there are no gaps. The joints of the cloth are stiched tightly using cords 1/2 nylong twine.

The results obtained are as follows:-

Total seed stocked	= 33.35 lakhs.	
Seed realised in 6 ponds	= 9.44 lakhs	
Survival (Average)	= 32.96 per cent	
Size of the seed	= 45–60 rnm.	
Out of which		
Rohu stocked	= 6,25,000	Per cent realised–60.42 per cent
Realised	= 4,02,920	
Mrigal stocked	= 27,15,000	–Do– = 20 per cent
Realised	= 5,42,120	

The entire experiment was conducted for a period of 3 weeks only. Stocking was done between 22.8.1983 and 5.9.1983.

The main constraints for obtaining better crop was:-

1. Use of in-organic and organic manure, food material and ammonia gas to control the weeds,
2. Submerged weeds (Ceratophylam, Chara mass growth of alqal bloom).
3. Absence of adequate food (Zooplankton).

There is considerable scope to improve over the management programme to get better production. From his experiment the author observed that the stocking density was very high and the use of organic and inorganic manure and ammonia gas should be used with caution.

Cages for Fish Culture

Floating net cages for fish culture in large reservoirs are often subjected to severe physical damage due to wave action. Mukherjee (1990) for the first time made some experiment and suggested flexible floating barriers, sheet or mattress barriers and rigid floats are some of the structures which have been tried with commendable success in protecting the aquaculture infrastructures both in inland and off-shore water bodies.

The floats dampen the wave thrusts and absorb the wave energy before the waves can propagate and strike the cage enclosure causing damage of any kind. The author opions that this system requires further study.

Earlier Kumariah and Parmeswaran (1985) proposed a circular cage that could be used in reservoir with, moderate wave action for culture of carps, tilapia and airbrathing fishes. The Cages can float at the bottom. Floating cages are considered to be the most appropriate for Indian condition and the experiments conducted so far in the country for seed rearing. Groundnut, nutrition and biomonitoring have been possible in such enclosures.

Singit *et al.* (1985) experimented rearing of Indian major carps in Tungabhadra Reservoir during 1984-85 in four floating cages. The cages were made of 16-P velon screen fitted on rectangular bamboo frame 10 x 4 x 1 m and stocked with Rohu and mrigal. The experiment was vitiated due to destruction of Cages by heavy winds. The results obtained at the end of 3 months was–survival rate varied between 37.5 and 87.5 per cent when stocking densities ranged from 2-5 m ha^{-1} when growth was about 100 ram (33g).

It is thus evident that number of fingerlings of suitable size produced in the pen from fry can be released into the reservoir according to their sizes. Fry cultured in nurseries generally are very delicate and to avoid any mortality by directly introducting them into the new environment of reservoir it is worth while to rare the fry of nurseries to a certain size in pens or cages and then released into the reservoir.

Cage Culture in Getalsud Reservoir

In the absence of nurseries and rearing ponds near the reservoir, can be reared in cages to fingerlings size before release into the reservoir. (Inland Fisheries Research/ Information series 4 on Gitalsud, Final Report 1971-81, March, 1984), CIFRI Barrackpore).

An experiment in cage was started in August 1974. The cages were fabricated of thick bamboo which is lighter than water and fixed near the dam with a view to minimise the wind and wave action on the cage, Synthetic floats were fixed to the cages and adjusted in such a manner that the cage remained submerged in water to a depth of 1.1 m to 1.2 m. The cages were anchored by sufficiently long ropes to allow them to rise up with increase in reservoir water level. Each cage was provided with an inner lining of synthetic 'hapa'. In the initial stages of rearing, hapas of 1/20 mesh were used but later as the size of fry increased hapas of 1/8" mesh were used for raising fingerlings.

Fry of major and medium carps were stocked in the cages at a maximum density of 2500/Cage (70,00000/ha) which is much above the conventional nurseries. The length of the fry at the time of stocking varied from 10.31 mm. In subsequent experiments the rate of stocking was reduced to 300-2200/cage (7,50,000–52, 50,000/ha).

During the first two months the average growth being 17 mm/month–*Cerrhinus mrigal*, 25 mm/month for *Catla catla* and 20 mm/month for *Labeo rohita*. In winter months the growth rate was 14" in *C. mirgals*, 30" in *Catla* and 17" in *rohu*, 19" in *labeo bata* and 22" in *Cirrhinus reba* per month.

MOC, GOC and RB in the ratio of 3:1:1 were used as supplementary feed. The young ones were fed at the rate of 30 per cent body weight for 3-4 days which was later reduced to 2 0 per cent. The feeding was then gradually reduced to 4 per cent during winter months and again increased to 10 per cent during summer when water temperature ranged between 28 and 34°C. Feeding was done twice daily (morning arid afternoon).

Production achieved during the year of rearing was 22.5 kg/Cage.

Chapter 21

Ecosystem vis-à-vis Reservoir Production

Reservoirs after their impoundment commences their trophic within 2-3 years when the vegetative areas with rich soil get flooded. The initial biological process are dominated by the microbiological utilization of the flooded organic matter with the help of large quantities of biogenic substances released from the soil and the organic matter after their decomposition. Of the entire reservoir shallow bays having high shape factor are the most productive zones. Such ecosystem has a very high amount of available energy which differs from Reservoir to reservoir. Further reservoirs in North India belt come under the category of monornictic lakes with strong thermal stratification during summer. But during winter both wind action and monsoon floods breaks the summer stratification (Rihand and Govind Sagar). In peninsular reservoirs, on the other hand generally show absence of any sharp thermal stratification (Nagarjuna Sagar and Bhavani Sagar) circulation of water brings locked chemical nutrients of the tropholytic zone upto trophogenic zone and facilitates fixation and utilization of energy. (Heat dynamics of 4 impoundments are shown in Figure 12). Reservoirs like Bhavani Sagar, Nagarjuna ar and Govind Sagar on the other hand have shown strong oxygen decline in the tropholytic layers and distribution of oxygen is klinograde in nature. These reservoirs can therefore be termed as high energy reserve at the bottom and as such come under the category of high productive reservoirs. On the other hand Rihand showed a near uniform distribution of oxygen from surface to bottom (Orthograde) and can be put under low productive reservoir, as bottom energy is of low order. This gives a clue how to segregate the productive reservoir from low productive one. Hence, if proper management is not done in the productive reservoirs it may lead to enormous population growth of weed-fishes which ultimately create a condition most suitable for growth of larger population of cat fishes at higher trophic levels. Once these cat fishes get a foot hold in the system, the management will be more difficult (discussed elsewhere). It is therefore essential to manipulate the system during the early stages so that vast amount of available energy is properly utilized by stocking with fishes feeding at lower trophic chain. Such phenomenon is witnessed in case of Nagarjuna. Sagar and Bhavanisagar reservoirs. In Nagarjuna Sagar cat fishes have established to such an extent that only 20 to 25 per cent of the available energy is utilized by primary consumers and 75 per cent by cat fishes (Secondary and tertiary) (Natarajan and Pathak, 1983). The reservoir shows poor energy harvest, as fish though limno-chemical features indicate its high productivity. In case of Bhavani Sagar where 68 per cent of the energy is utilized through primary consumers, the energy harvest is far better than that of Nagarjuna Sagar.

Natarajan and Pathak (1983) after their extensive investigation on pattern of energy flow in some tropical and subtropical impoundments concluded that adequate management of reservoirs well in time is the only solution to have a better fish production.

Individually none of the factors stated earlier (Tables 30 and 31) appears to have any relation with fish production, but collectively they are related and above all it is the management that brings the success. This is why probably Dubey (MS) has given more stress on proper stocking. Natarajan and Pathak (1983) from their study have shown that catfish are abundant in Nagarjuna Sagar and available energy consumed by cat fish is 75 per cent where as primary consumers utilize only 20-25 per cent, thereby affecting low production, where as in the case of Bhavani Sagar although 68 per cent energy is utilized only Catla fishery dominates, but still proper stocking with multiple species at shorter food chain will improve the yield. Rihand reservoir on the other hand has maxium energy reserve in comparison to other reservoirs. Its proper utilization by stocking with multiple species at the shorter food chain will improve the yield. Rihand reservoir on the other hand has maximum energy reserve in comparison to other reservoirs. Its proper utilization by stocking with multiple species at the shorter food chain will improve the yield. Such findings are of vital importance in management programme.

Incidentally it may be mentioned after determing the status of reservoir energy, in the second phase the stocking may be initiated only after revealing the individual stocked position of fishes in the reservoir on the lines suggested by Jhingran and Natarajan (1969) during their study of DVC reservoirs. Both the studies are intimately related.

Since natural stocking is uncertain and insufficient adequate stocking as suggested by Dubey (MS) in small, medium and large reservoirs must be done so that the entire food is consumed to have a healthy growth and reproduction of fish.

Morphoecological conditions and productive potentiality of some reservoirs are shown in Table 31.

Table 31: Morphoecological Conditions and Productive Potentiality of Some Reservoirs

Year of Commission	Reservoir	Area FRL (ha)	State	Shore Development	Vob Development	Water Level Fluctuation (m)	Mean Depth (m)	Wind Velocity	Fish Production			Source
									t	kg/ha		
1957	Hirakud	71963	Orissa	13.5	0.52	–	11.3		350	4.8	1985–91	Sugunan (1995)
1960	Gandhi Sagar	66000	Madhya Pradesh	4.78	0.601	50.78	11.73		738	11.1	1979–81	Dubey (MS)
1962	Rihand	46538	Madhya Pradesh	7.04	0.89	15.2	22.8	Highest June–July	167.29	5.707	1971–81	Sugunan (1995)
1953	Tungabhadra	37814	Karnataka	5.251	0.737		9.8		3085	81.5	1980–81	Sugunan (1995)
1967	Nagarjuna Sagar	28427	Andhra Pradesh	7.89	1.13	72.3 ft.	40.5	May–Oct. 28 km.	118.79	6.63	1971–80	Final Report (1983 March)
1974/5	Ponds	24539	Himachal Pradesh	2.48	–		35.7		484.8	32.3	1976–90	Sugunan (1995)
1963	Govind Sagar	16867	Himachal Pradesh	12.86	1.01	50.78–1988	55		556.91	55.69	1970–93	Sugunan (1995)
1934	Mettur (Stanley)	15346	Tamil Nadu	6.7	1.4	43.25–1989–90	18.3		344	22.76	1967–77	Sugunan (1995)
1932	Krishnaraj Sagar	13200	Karnataka	7.25	0.78	13.32	30.17		161.93	12.27	1980–88	Sugunan (1995)
1955	Konar	2590	Bihar	8.78	0.85		12.97		11.785	6.408	1967–68	Jhingran and Natarajan (1969)
1952	Tilaiya	5921	Bihar	9.1	0.766	–	6.65		23.999	6.451	1985–91	Jhingran and Natarajan (1969)

Contd...

Table 31–Contd...

Year of Commission	*Reservoir*	*Area FRL (ha)*	*State*	*Shore Development*	*Vob Development*	*Water Level Fluctuation (m)*	*Mean Depth (m)*	*Wind Velocity*	*Fish Production*			*Source*
									t	*kg/ha*		
1953	Bhavani Sagar	7876	Tamil Nadu	4.0	1.0	5.46– 1979–80 22.06– 1977–78	11.5	Jun–Aug. 24–40 kg.hr.	193.488	24.5	1971–81	Debraj *et al.* (1990) Seminar
1981	Haemavathy	9162	Karnataka	1.6	0.754				130	13.65	1984–86	Sugunan (1995)
1978	Ravi Sankar Sagar	9540	Madhya Pradesh	2.96	2.0	7.14	9.50		31.35	4.91	1987–93	Sugunan (1995)
1963	Ramgarh	1260	Rajasthan	1.81	0.61		4.58		132	104	1979–81	Dubey (MS)
1957	Santanur	2010	Tamil Nadu	6.2	1.0		10.1		128.638	64.0	1966–84	Debraj *et al.* (1990)
1958	Amarvathy	850	Tamil Nadu	2.3	1.2		13.7		71.520	131	1961–84	–Do–
–	Sandynulla	258	Tamil Nadu	4.5	1.2		10.3			45.4		Sreenivasana (1978)
–	Tirumurthy	466	Tamil Nadu	–	1.0		11.0		13.0	–		Sugunan (1995)
1971	Gitalsud	3459	Bihar	3.85	0.6839		32.6		3.867	–		Final Report 1971–81 Ecology and Fisheries of Freshwater Reservoir March–1984

Shallow bays rich in ecological niches are supposed to be productive and such areas are to be delineated and protected from the main reservoir.

Dubey (1969) has discussed the productive zone of some developed reservoirs of Madhya Pradesh (Table 14).

According to Dubey (1969) in seven reservoirs studied 33.3-95.9 per cent of the area are productive zones which yields between 45.3 and 203 kg/ha (average 41.1 kg/ha) of fish against the total production 15.1-153.7 kg/ha (average 76.4 kg/ha) at FRL. Dubey (MS) has further observed that by adopting different techniques large and very large reservoirs too can be made productive. This he observed in case of Govind Sagar (16,867 ha) were yield of 200 kg/ha/yr was attained, 220 kg/ha/yr in case of Ukai (52,000 ha) and 159 kg/ha/yr in case of Bhavani Sagar (7876 ha) against their earlier production as 55.69 kg and subsequently 70.30 kg/ha/yr, 570 kg/ha/yr and 63 kg/ha/yr respectively.

Thus it is clear that ultimate object of the findings of Natarajan and Pathak (1983) and Dubey (MS) is to stock the reservoir adequately with proper composition and size and irradication of predatory and other weed fishes that compete for food from economically important spicies to get maxium production.

Jhingran and Tripathy (1969) has given an account of the population of predatory fishes in some reservoirs of India which is quite alarming. According to them *Mystas* spp. *Channa* spp. *Wallago attu, Silonia silondia* and others dominates the catches of many reservoirs. A study of the catch-composition of some reservoirs in Uttar Pradesh has shown that catfishes contributed 23.30 percent and 36.8 percent in Matatila (Jhansi district) and Sarda Sagar (Pilibhit district) during 1963-64; 39.79 per cent in Sarda Sagar during 1964-65; 27.20 per cent in Chandra Prabha, Nangarh, Bhainsura and Latifshah (Varanasi district) and 60.94 per cent in Sarda Sagar and Nanaksagar during 1965-66. Similarly in Tungabhadra reservoir (Karnataka) cat fishes dominate both by numbers and weight, Rajan (1969) stated–a 711 mm long wallago is able to consume carp measuring 230 mm in length. Besides these catfish problem, reptiles, birds and mammals are also present. Sirsi and lower Khajuri (Mirzapur district), Nanaksagar (Nainital district) and Sardasagar (Pilobhit district) reservoirs in Uttar Pradesh has large population of crocodiles. All these need effective control for building up a population of commercially important fishes feeding on lower food chains.

The study of energy flow in reservoirs as done by Natarajan and Pathak (1983) gives important informations of the productive potentiality of a reservoir prior to stocking so that stocking can be initiated accordingly, while CIFRI based on the extensive study during 1971-1985 on the projects "Ecology and fisheries of seven freshwater reservoirs" have shown importance of various aspects of reservoirs such a productivity and characterization of reservoirs, fish food organisms stock characteristics, food fishes including their breeding, recruitment, size of gillnets and management. By adopting different technologies it has been observed that the production in large and very large reservoirs too can be achieve as high as 200 kg/ha/yr in Govind Sagar, 220 kg/ha/yr in Ukai and 159 kg/ha/yr. in Bhavani Sagar.

It is thus clear that irrespective of size and region, stocking and management is the only solution. Further in India most of the reservoirs are quite old, mostly commissioned from 1950 onwards, and have attained sufficient rich ecological nishes with some variations depending upon the nature of fall in the water level. This is a common phenomenon, as considerable organic material are brought into the reservoir along with the silt load and thus the richness is maintained. At the moment it is the understocking and improper management that leads to low production. Dubey (1969) opioned that it is possible to achieve higher level of production in the range of 75-120 kg/ha/yr in the average category of reservoirs and 250-500 kr/ha/yr. in small and medium reservoirs. Thus it is evident that if all the reservoirs are scientifically managed as highlighted above and data properly presented the yield per hectare will definitely improve considerably and will be much higher than presented elsewhere (Table 27).

High silt precipitation in reservoirs (Tables 6 and 7) which is quite considerable (annual rate of silt precipitation is several times more than assumed silt load) reduces the water holding capacity, may not have immediate effect on fish fauna, but in the long run due to progressive change in the morphoecological condition (when at a later stage it will turn into a lake) may have impact on aquatic biomass including fish population which requires periodic study. The dredging is thus vital to save this dead storage and other areas.

In case of drought or atmospheric high temperature (Global warming) problem arises for fish culture in reservoirs as considerable volume of water has to be discharged for irrigation and other purposes, and thereby minimum water level can not be maintained in the reservoir. Such case happened in case of Krishnaraj Sagar reservoir of Karnataka in 2002. "Faced with its worst drought in 30 years, the South West Monsoon has brought 42 per cent less rainfall this year (September, 2002) in Krishnaraj Sagar. Already one fourth of the reservoir has gone dry as the daily flow of the water into it is 948 causes against an outflow 3,806 cusech, and if this trend continues the dam will be without water in less than a month's time." (Ghosh–2002) Added to it on some occasion of summer, sudden high rise of atmospheric temperatures (perhaps due to the Global warmings) increases the water temperature also where fish fauna finds inconvenience to survive.

Under such circumstances there will be terrible setback both for fisheries and agriculture. The dead storage level of reservoir has to be maintained properly to save the fishery, and for agriculture, adequate deep tubewell system etc. should be thought of as is normally done.

At the time of impounding streams or rivers no attention is generally paid to the clearance of submerged obstruction from the river bed which has a vital importance to facilitate easy fishing. As a result the reservoirs misleads the actual production against their stocking. The magnitude of the problem would become clear from the fact that 629 houses in Tilaiya, 118 in Konar, 1583 in Maithan and 2119 in Panchet hill reservoirs and 90 villages and 366 towns and villages in Govind Sagar lake have been submerged apart from large forest tracts, rocks and houlders etc. (Jhingran and Tripathi, 1969).

Finally expression of production figure is not clearly stated, whether production shown is based on full reservoir level or average area etc. IIMA (1983) have suggested that because of high fluctuating level of water, the area for the purpose of production per/ha has been taken as 60 per cent of the FRL, while National Commission Agriculture (NCA) has suggested that for uniformity, the average area between FRL and DSL should be considered for calculating the yield/ha.

Chapter 22

Fishing Nets and Gears

Experiments on new fishing techniques are conducted with a view to improve the existing fishing methods. In reservoirs the development of modern fishing received a boost with the introduction of synthetic fibres for fishing. These fibres have virtually replaced the cotton and other natural fibres. Though our country has a number of artificial impoundments with a well developed fishery, comparatively very little work has been done regarding introduction of new techniques of fishing. Gulbadamov (1962) suggested improvements on the years used based on his work at Metthur, Krishnaraj sagar and D.V.C. reservoirs. Tripathy and Krishnamurthy (1964) reported experimental fishing observations at Mettur dam. Krishnamurthy, Govind and Bhatnagar (1964) described experimental fishing with gillnets in Tungabhadra reservoir. Znamensky (1967) suggested that the ineffective gear should gradually be replaced by nets of improved design. He also suggested formulae to determine the mesh size of nets in relation to length and weight of fish sought to be captured and strength of twine.

The aim of experimental fishing is to evolve suitable technique for a rational exploitation of the fishery. The effectiveness or efficacy of a particular net is gauged by the relative quantity of fish it captures to another gear. Von Brandt (1955) indicated the difficulties in defining efficiency and suggested that efficiency should be measured in terms of relative abundance of number of fish caught or weight of fish obtained.

In most of the reservoirs of the country use of gill net is common. It is also necessary that trashfish should also be exploited as these uneconomic species compete for food with the commercial important fishes. In order to obtain a high productivity these should be prevented from gaining stronghold in the reservoir.

In most of the states in North India wooden boats with or without keels are used in reservoirs, operated generally by two fishermen, where as in South India, a special type of craft known as 'CORACLES' are commonly used for catching fish. In certain large and very large reservoirs mechanised boats are used as mother ships for toeing the fishing boats to distant fishing zones and carrying fish to the landing centres.

Various fishing devices have been tried in order to perfect the tackle suited to the reservoir fisheries and their efficiency assessed. Various nets used for fishing operation by different workers in different reservoirs are indicated below. F.A.O. experts surveyed some storage reservoirs in India and suggested some improvements and modifications of existing reservoir fishing methods (Gulbadamov, 1961 and 1962). The improved methods of experimental fishing was conducted by the Lacustrine unit and the success of operation of the new designs are dealt by Krishnamurthy *et al.*, 1964) and further detailed observations were made in Methur reservoir by Ranganathan and Natarajan (1969) and by David and Rajagopal (1969) followed by Krishnamurthy (1969) in their study of Tungabhadra Reservoir.

Ranganathan V. and Natarajan, V. (1969)

Various Nets used in Mettur reservoirs are indicated below:

Shore Net

This net is made of fairly strong cotton thread circumference of the open end is 44 ft. mesh 1 square and height 12". circumference is fringed with lead. A ring is attached towards the closed end through which the hanging rope runs freely. These rope divides into a number of other ropes, of smaller thickness. These ropes are attached to the border of the opening like ribs of an umbrella. Each of these subsidiary ropes bifurcates into two before it is fixed up to the margin of the opening. While casting, the hauling rope strips out through the ring and the net spreads out in the water and settles down quickly because of the weights. The net is pulled up when the hauling rope is dragged up. When the hauling rope is pulled out all the lead weight gather towards the centre and the ends are turned upward, thus closing the rough mouth of the net, imprisoning all the fish within. This is effective in shallow waters. It could also be made use of in deep waters. The net was tried, but found unsuitable during deep water fishing experiments in the reservoir.

Perch Traps

Chiefly used for capturing eels, catfish etc., which damage the ordinary nets. Crabs or small fish are used as bait. These are made of bamboo splinters or iron bars. These traps are placed at the bottom of the water or suspended at mid-column.

Sarattu Valai

Used to catch big fish. This is a bigger cast net 5 yeards in length, about 13 ft. in radius with 18 ft central rope and 1" mesh plied from mid-waters.

Wall Net (Udu Valai)

Anchor or wooden peg is driven into the earth. Floats of small bits of hollow reed or floats of catatropis strung through the rope, sinkars are of burnt earthern balls passing through the cord made of common coarse hemp, Three pieces each of 100 ft, breadth $2^1/_2$ ft. mesh 2" sq. and also $1^3/_4$ sq. They are of two types. One end of the net is fixed to the shore by a sharp pointed wooden peg tied to it. Then the fishermen get into the water by a 'Caracle' and one paddles away simply plying it into deep waters as they proceed. After the net is cast, it slowly settles to the bottom and it sinks because of the bloats unable to sustain the weight of the earthen balls below. After it settles down, it stands erect in the form of a curtain by the help of the floats. Only the bottom feeders are caught by this net, Extensively used in deep water fishing (100 to 200 ft. length).

Oosa Valai

Improved udu Valai, Cotton, $2^1/_2$" mesh, drift net. Fish are gilled.

Malabar Hemp Drift Net

Of equal measurement as Udu Valai, oblong floats of 6 to 10" long and $1^1/_2$" sq. across are used as weight. The number and frequency depend upon the depth at which it is operated.

Rangoon Net

16 pieces joined at the ends. Each piece is 50 ft. long. This stands like a screen with floats overhead made of wooden pipes. No weight is used. Hence it can drift along with the current. Mesh generally is 2" square. Very fine cotton of 20 counts is used. One end attached and the other end left afloat. Most successful in deep water fishing.

Nalu Vali (Raju Valai)

Oue piece only 150 ft. This net is another type of wall net, having differences in make and length. It is taken across shallow areas and two pipes are driven into the soil. It stands errect in the form of a screen, usually the two pipes at either side sweep the net in the water when the fish are entangled. They may also close up forming a circle with the net enclosing the fish. When fishes try to escape out of the cordon, they get enmeshed. This is not used in deep waters. This net is 6 ft. deep and 50 to 100 ft. long Burnt clay balls of about 1" in diameter are used as sinkers and cylindrical floats make the hollow light wood (3-4" long) are also used. The net is having $1^1/_2$" mesh and 1 mm diameter thread (12 ply).

Night Lines

This contrivance is used to capture predators like Wallago attu, murrels etc. Live frog is used as bait. The frog is bloated by means of straw tube introduced in its anus and blown up. The bloated frog do not sink down, rather kept in position, inviting the fish.

Long Lines (Pandam)

This consists of 50 to 100 hooks attached to strong thread. Small fish, etc. are put as baits and both ends are tied to the shore.

The final and successful resort is found in Rangoon nets for wide spread application, since the other tackles did not offer successful results. From the year 1949 onwards, an intensive exploitation started making use of Rangoon nets, firstly with the parachute cord silk and then tied with synthetic yarn of various codes. Mesh sizes of these nets were regulated to suit the sizes of the different fishes. The efficiency of the 'Mettur Rangoon' nets in Bhavani Sagar reservoir is discussed by Ranganathan and Venkataswamy (1967).

In the initial stages, fishing was carried out with Udu Valai, Malabar hemp drift nets, Oosa Valai, Crabpots, Perch traps, long lines, Hook and lines, Rangoon nets and Catla nets. In later years, Rangoon nets with varying mesh sizes and catle nets came to the forefront owing to their efficiency over the other fishing tackles.

David A and Rajagopal K.V. (1969)

Detailed studies on experimental and commercial fishing in Tungabhadra Reservoir are indicated below:

1. Experimental Fishing methods:
   (*a*) Sub-surface gill nets,
   (*b*) Bottom Gill nets,
   (*c*) 'Hobbled' Gill nets.

Sub-surface and Bottom Gill Nets

Both in the case of sub-surface and bottom set gill nets, 3 to 4 nets (30 m. hung length of each) of a mesh range of 30 to 150 mm were laced together to form one unit of 90 to 130 m in length by 5 ra deep when hung. The same set of nets could be used as bottom set nets by increasing the number of sinkers and decreasing the floats proportionately and using desired lengths of anchoring ropes.

These net units were transported in a coracle to the fishing site add a stone auchor lowered with the master float at one end and the net was laid either parallel, perpendicular or diagonal to the shore, depending upon the water current and direction of wind. When the net was completely laid, the second anchor was lowered and the second master float laid in position at the other end. When there was strong water current the anchoring ropes were increased to twice the depth of water. In case of surface net, the number of sinkers were slightly increased to avoid rolling of nets due to current and wind action. The catches were removed every 24 hours usually in the early morning hours. The nets were taken out for repairs and change in fishing position usually once in 4 days.

'Hobbled' Gill Nets

'Hobbled' gill nets were used in the TVA reservoirs and in some N.American reservoirs and are specially designed to catch flat-head catfishes. In 1966 the authors fabricated the 'hobbled' gill nets on lines described by Heard, R. William (1959) and introduced it in the Tungabhadra reservoir.

In this, net vertical 'hobble' lines run through the webbing of a normal gill nets and are tied to float and bottom lines to prevent the webbing from hanging its full vertical depth. The bobble lines are tied at intervals of 1.8 ra at each float with 0.6 m hanging drop (attachment of webbing to float and lines) in between the two floats.

This 0.6 m hanging drops allows the loose webbing to slide (20 meshes in 30 mm bar, 18 in 40 mm, 16 in 50 mm, 15 in 60 mm, 14 in 70 mm bar and so on) horizontally a distance of 0.6 m. The horizontal looseness in the webbing obtained from 0.6 m, hanging drop, aids the webbing efficiency of the net and allows most fish to get caught without enlarging in the 0.6 m linear section of the net. The horizontal hanging co-efficient of webbing here also was 0.5. The net was operated in reverie and transitional zones of the reservoir where lotic conditions prevailed. Hobbled gill net is essentially a bottom set gill net, operated 25 m, away from shore and perpendicular to shore in depths of 5 to 10 m.

Commercial Fishing Methods

The fishing gear of commercial fisherman in the Tungabhadra Reservoir were varied. There was no standardiation with regard to mesh sizes, length and breadth of the nets or code numbers of nylon twine used for webbing. However, the hanging co-efficient of 0.5 was being maintained by almost all the reservoir fisherman. Though mesh sizes varied from 30 to 150 mm, mostly fishermen employed 30 to 60 mm. mesh bar nets irrespective of any limits to mesh size of numbers of nets as their object was to catch as much fish as possible without any consideration towards conservation of breeders. The few different types of gears used by fishermen in the Tungabhadra reservoir were as follows:

1. Rangoon Net ("Catla" nets),
2. Uduvala (Wall net),
3. Pedda–Uduvala (Wall net)
4. "Alivi" (Shore Seine),
5. Kontevala (Drag net),
6. Cast Net,
7. "Betsuvala" or 'Chata' net,
8. Long lines.

Rangoon Nets

Gill nets known as Rangoon nets are generally employed for subsurface fishing. The nets however are not used as drift nets as in the case of original 'Rangoon' nets common in the Godavari–Krishna delta. The larger meshed nets from 90 to 150 mm. bar are known as 'Catla' nets as, larger fishes like *Catla catla* are ejected to be captured by them. These nets as in vogue on the Tungabhadra reservoir were between 80 and 100 m. in length and 5 to 7 m in height.

The nets were being operated almost throughout the year but the transitional and shallow zones of the reservoir were being well exploited by employing these sub-surface gill nets. Inundated bays and main reservoir region (Deep zone) were served by these nets where water was sufficiently deep and free from submerged tree trunks generally in the post monsoon months when some stray large fishes were being caught.

'Uduvala' or 'Bidubale'

These are bottom set gill-cum-wall nets with mesh sizes ranging between 35 to 80 mm.(bar) with lengths of 60 to 90 m and widths 0.50 m to 0.75 m. The head and bottom ropes pass through each mesh and clove hitched by reed floats to every third mesh in the head rope with stringing of burnt clay oval sinkers to each mesh in the bottom rope. This attachment of mesh directly with head and bottom ropes gives a firm wall like tension to the net. $1^1/_2$ mesh selvedge all round is provided for the net. These nets are efficient and are generally operated in creeks and bays in South Indian reservoirs especially by Metthur fishermen and Killekethas in Mysore reservoirs. They are operated in upto 3.0 m depth parallel or diagonal to the shore being laid in the evening and hauled In the morning,

Pedda-Uduvala

This is a modified bottom gill net possessing advantages of both the 'Rangoon' and Uduvala nets, as these nets are wider (about 1.5 to 1.75 m) which increases the surface area. They are operated in deep areas of the reservoirs. Construction details are similar to uluvala.

Alivi (Shore Seine)

The Alivi (Shore Seine) is 75 to 100 m in length with widths from 3 m, at the outer edges to 6 m in the middle. The net is made of cotton twine and tanned in vegetable tanning solution. The tapering edges act as wings, the meshes gradually decreasing in size from 25 mm. (bar) to 20 and 12 mm on both sides but as little as 7 mm, in middle for a length of about 45 m. Hence, the middle position acts as an effective seive net. Coir ropes are tied as head and foot ropes and continued on the outer side as a single rope for 100 m on both sides, A bamboo is tied on the two outer edges for rigidity. The head rope is provided with 70 to 75 wooden floats of indigenous material at intervals of 0,75 m. to 1.75 m. The 'alive as usually operated in the sluggish river streams in the lower Godavary and the Krishna is now adopted by fishermen to the inundated areas of the Tungabhadra reservoir Two coracles and eight men are required to operate one 'alive'. The nets were operated on the Tungabhadra reservoir at all hours but generally from mid-night to early morning hours for a duration of 8 hours which can be termed as being the unit of 'fishing effort' for the net. The entire operation of 'alive' consists of (*i*) encircling, (*ii*) Scaring fish particularly by outer ropes to the middle and (*iii*) effective seining.

No fish can escape once the net part encircles an area of the reservoir; a good idea of the standing crop in the zonal areas can be obtained by studying the catch (David, 1969).

Kontevala

This is an almost miniature 'alive' in shape and appearance. Its total length is 20 to 25 m. width from 2.5 to 3 m. in the centre, tapering on to two edges to 1 to 1.5 m. Cotton twine is used as in 'alivi' and meshes are uniformly 10 mm (bar) throughout. Bamboo sticks of 0.75 m are tied at intervals making several pouches on the net during operation. The net is usually dragged by 3 men in littoral areas and the catch comprised of young fish and prawns, mainly used as bait for long lines.

Cast Net

Cast nets are not effective in reservoirs. Rarely they are operated in reservoirs to capture small fishes–mainly minnows and trash fish. They were not employed in the Tungabhadra reservoir.

'Betsuvala' or 'Chela' Net

A small sub-surface net was used for capturing minnows by Killekethas. A fine meshed 10 mm. (bar), cotton twine net (0.50 to 0.75 m wide and 10 to 20 m long) was operated in margins and littoral areas of the reservoir. The net was provided with small floats (reeds) or broken dried gourds, and burnt clay sinkers. Mainly (Oxygaster spp; *Puntius* spp., Aspidoparia morar Ambasis ranga etc. were being obtained: The net was useful mainly at the time of reservoir's high water level during monsoon months.

Long Lines

The long lines consisted of a single line having vertical lines (snoods) with barbed hooks of preferably small sizes (number 10 and 12). The snoods were so fixed that they could not be entangled with one another. The length of snoods were of uniform lengths or variable. The lines with usual baits (earthworns, young fish, prawns etc.) could be operated on surface in column and bottom as required, commercial fishermen operated the lines mainly on the reservoir surface. However, the following modes of operation are useful in reservoirs.

1. Surface long lines with 300 to 500 hooks with floats tied at distance of 5 m.
2. In column line, the floats are tied at 10 m. intervals,
3. In case of bottom long line, no floats are used but very small floats are tied behind each hook as a buoy. The main line rests on the bottom (Gulbadamov, 1962).

Two sinkers of sufficient weight and, two master floats on either end of the main line are provided in all three months.

In comparatively deep storage reservoirs where fish movements are fouund at all depths, oblique long lines may be tried. This is a simple long line with snoods of 0.5 m length attached at an interval of 1.5 m, an appropriate anchor at one end and with adequate floats at the other. Care has to be taken to maintain the oblique long line with a range in angle of 30° to 60° to prevent the snoods resting on the main line.

The following nets which are not commonly operated in lakes and reservoirs may be tried in storage reservoirs in India after suitable trials and modifications:

1. Fyke net.
2 Frame net.
3. Multimeshed gill nets.

Fyke Net

It is a small bag net mounted on 3 or 4 circular hoops; these hoops help the net from collapsing and form the attachment base for the conical entrances which prevent the fish from escaping. The catch is removed from the last pouch through a collapsible opening. The fyke usually has short, vertical, wings obliquely on either side of the opening to guide the fish into the first compartment. Fykes are used in shallow zones of the reservoir especially to trap the bottom feeding fishes. If oblique wings are used the net has to be set with auchors, and wings should be provided with heavy sinkers on the bottom and floats on the top (Rounsfell and Everhart, 1953).

Frame Net

The principle is the same as in the 'Rangoon nets[1], the only modification is that both vertical and horizontal lines pass through the meshes at 1 m intervals (1 m square) where webbing is loosely hung throughout its length and width forming a pouch. In case of damage to the net by fish, only that particular square gets damaged as the vertical and horizontal lines prevent strain on the neighbouring square. The net is generally operated in riverine and transitional zones of the reservoir where some fluviatile conditions prevail. The looseness of the net offense ressiliance to the current and chances for fish to get entangled becomes greater.

Multimeshed Gill Nets

Type I. The principle of construction of the net is the same as Surface or bottom gill nets, except for some modifications. The different mesh sizes vary vertically throughout their lengths. The net of 60 m. long (hung) will thus consist of meshes 70 mm. to 30 mm. (bar) from head/rope towards the centre and 70 mm to 40 mm from bottom rope towards the centre.

Type-II. Net of 60 m long (hung) having mesh sizes 70 mm to 30 mm (bar) from head rope to bottom rope braided vertically throughout its length.

Multimeshed gill nets are not operated in India at present even on an experimental scale.

Experimental Fishing in Tungabhadra Reservoir

The experimental nets, namely, sub-surface, bottom and hobbled gill nets operated during April to December 1966 within the Tungabhadra Reservoir, provided an idea as to the efficiency of each type of net. Experiemental fishing was conducted for 1830 net days and a total of 631.905 kg of fish were caught in all the three types of nets. In general, tine Catch/net day was highest (0.386 kg), for bottom set nets followed by hubbled gill nets (0.389 kg) and sub-surface set nets (0.297 kg).

The carps predominated (64.4 per cent to 84.4 per cent). Over cat fishes in the sub-surface nets in all the zones of the reservoir during April to December, 1966. During July to September, they were poorly represented in riverine (10.9 per cent) and deep zones (72.2 per cent). The hobbled nets also showed predominance of carps over cat fishes in all the zones during October to December. (Riverine zone 92.3 per cent, Transitional zone, 71.7 per cent, shallow zone 80.3 per cent and Deep zone, 58.2 per cent), but during July to September, carps formed only 38.4 per cent of the catches in deep zone.

In all experimental gill net catches, the carps in the order of dominance were represented by–*Puntius kolus, Labeo fimbriatus, Puntius Sarana, Labeo Calbasu*, and *Osteobrama vigorsii*. The Catch comprised of *Silonia Childrenii, Wallago attu, Rita pavimenta, Mystus aor and Mystus seenghola.*

Commercial Fishing in Tungabhadra Reservoir

Observations were made in detail on the landings of the 'alivi' nets (shore seines) between 1964 and 1966. This gill net catches could not be observed regularly as units were distributed unevenly all through the year.

The 'alive' net is generally operated during low water level of the reservoir *i.e.* between January and June in the riverine, transitional and shallow bays, an average 'alivi' with a length of about 300 m (including 100 m scare ropes on each end) covers approximately a semi circular area of 1.4 ha. Standing stocks at the 3 zones of operation were estimated to be 6.441 kg/ha to 20.453 kg/ha during the three years of studies between January and May; representative samples of all available species could be removed by 'alive' and gave the true composition of fish species and sizes available in the reservoir. Generally catfishes (75.8 per cent) dominated over carps (20.2 per cent), Among carps *Osteobrama Vigorsii* accounted for 10.1 per cent of the total catch next to which in order of abundance were forms like *Cirrhina reba* (3.5 per cent), *Puntius sarana* (3.3 per cent) *P. Kolus* (1.9 per cent) etc.

Mystus Cavasius (21.5 per cent), *M. aor* (12.4 per cent), *Wallago attu* (10.7 per cent), *Silonia Childrenii* (10.7 per cent), *Mystus seenghala* (5.8 per cent) and *Pseudeutropius toakree* (5.1 per cent) were of Commercial importance among catfishes.

The main species caught in *Uduvala* and *Pedda–Uduvala* were *Puntius colus, P.sarang, Osteobrama vigorsii. Labeo fimbriatus, Pseudeutro-plus toakree, Silonia childrenii, Mystus aor* and *M. seenghala.*

Experimental fishing in the Tungabhadra Reservoir for assessing the efficiency of different types of nets indicated bottom set nets to be more efficaceous (0.386 kg/net day) than hobbled gill nets (0.383 kg/net day) and sub-surface set nets (0.297 kg/net day).

It is observed that fill nets with mesh sizes 30, 40, 50, 60 and 70 mm. bar are more effective for carps in reservoirs and based on the observations made in Tungbhadra and other reservoirs, it is suggested that 30, 40, 50 mm. mesh bar nets may be used for capturing medium sized fish. 40, 50, 60 mm, for moderate sized fish and 50, 60, 70 mm. for large sized fish.

Gill nets are highly selective, as to species and sizes, the shore seine known as 'Alivi' in the Tungabhadra Reservoir and its related small dragnet, the Kontevela are known to be non-selective and provide data not only on standing stock of fish but also can eliminate undesirable, 'trash' fish and young of predaceous fishes on a large scale.

C. Krishnamurty K.N. (1969)

The various types of gear commercially used in reservoirs, the improved techniques, comparative efficiency of different ypes of gear and their commercial application etc. are discussed below:

Commercial Fishing

In view of the submerged obstructions, fishing activity is relatively of passive nature. At Mettur Dam where a well developed Commercial fishery exists, the fishermen use two types of gill nets, namely Rangoon net and Uduvalai. These nets have been popularised in other parts of the country and in many of the reservoirs, namely, Bhavani Sagar, Hirakud, Nizam Sagar, Krishnaraj Sagar, Tungabhadra and DVC, these form the main Commercial gear. Besides these, long lines are also used for the capture of catfishes. Cotton nets are completely given up and nets made of nylon or terylene form the mainstay of the fishing tackle.

The main gear used in reservoirs is the gill net which consists of a single wall of webbing which is held in vertical position with the help of a float line at the top and a lead line at the bottom. Floats and sinkers support an equal extension of the webbing. These nets can be used at surface, mid water or bottom by adjusting the floats and sinkers and the fastening lines and ropes.

Rangoon Nets

These are drift gill nets made of nylon Terylene. The length of the net varies from 25 to 30 metres and the width 3-4 metres. Mesh size varies from 25 to 30 metres and the width 3-4 metres. Mesh size varies from 50 to 150 mm. The design and construction of this net is very simple. The webbing is hung on to a head rope of a slightly thicker twine. Lead lines, breast lines and selvedge meshes are not provided. The hang-in co-efficient varies from 60 to 80 per cent. 10 to 20 net pieces of this type are conjoined and operated as a single fleet. Floats and sinkers are not braided into the net. Sometimes on the float line, small wooden pieces are attached which act as floats. The fleet is joined to thick coir ropes on either side, to which master floats are attached. To keep the nets in stationary position, master sinkers (stones) are tied to the bottom of the coir rope. Hence, the fleet of nets is kept in one area and the individual net pieces which are conjoined at the top, keep drifting, before and after for very short distances. The individual net pieces differ only in mash sizes ranging from 50 to 150 mm. The large mushed nets from 100 to 150 mm. are called 'Catla' nets, and big sized Catla are caught in these nets. The nets are operated with the help of a *Coracle* made of bamboo frame work and buffalo hide. Usually the nets are shot in the evening and hauled during the early hours of the next morning. After employing the cables, the nets are left back in the same area. Once a week or depending upon the catch, the fishing is grounded and is changed.

Uduvalai

These are bottom-set gill nets and only 0.5 to 1 metre in depth while the length varies from 50 to 100 metres. The webbing is made of either nylon or terylene and on the float line reeds are used as floats and mud sinkers are used in the lead line. Both these are braided into the net. A long coir rope or cotton rope is attached to the float line at the end of which a small wooden piece is attached which floats above the surface of water by which the net is located. The net is set at the bottom and the mesh size varies from 50 to 70 mm. Two or three nets of this type are operated at a time. The principal fishes caught by these nets are carps and catfishes. These nets are quite effective for a variety of fishes which are caught by getting enmeshed, entangled or by both.

Besides these nets, long lines are used for the capture of catfishes. Baited hooks of various sizes at regular intervals are hung with the help of snood lines on to the main line made of cotton rope. Cast nets, traps and to a certain extent, shore seines are also used in many of the reservoirs. But the catches are poor consisting mainly of minnows, mirror carps and sometimes young ones of major corps.

Experimental Fishing

Gill Nets

As gill nets form the most important gear, detailed work was undertaken at Mettur by Gulabadamov (1962). He found that the indigenous nets were defective in construction, hanging etc. To overcome shortcomings of the existing nets, new nets made of terylene were introduced and they had the following features:

1. Hang-in co-efficient was 50 per cent *i.e.* 100 metres of webbing was hung on to 50 metres of float and lead lines.
2. The vertical co-efficiȩnt of hanging was made into 0.85 *i.e.* if the stretched length of the webbing is 10 metres and the co-efficient is 0, 8.5, the resultant width of net is 10 x 0.85 = 8.5 metres.
3. Selvedge of two meshes width all round the webbing with a thicker twine than that of the main webbing was provided.
4. Breast lines were provided at the sides and lead lines were provided at the bottom.
5. Floats and sinkers were attached to the nets.

Though the efficiency of the nets increased with the provision of lead lines, breast lines and selvedge meshes, the presence of floats and sinkers hampered the fishing operations as these would get entangled in the meshes of nets. Krishnamurthy *et al.* (1964) used spherical plastic floats along the float line at regular intervals which could be removed easily.

By the weight of the lead like itself, the nets were held in upright position eliminating the use of sinkers.

Frame Nets

These are essentially similar to the gill nets mentioned above. The difference lies in the fact that the vertical hang-in co-efficient is reduced to 0.5 or 50 per cent *i.e.* 10 metres of webbing is reduced to 5 metres by providing additional vertical lines which run from the float line to the lead line at one metre intervals through the meshes. Additionally, along the horizontal position, lines are fitted at intervals of one metre from the float line to the lead line. The meeting points of both the vertical and horizontal lines were joined together and to the webbing. The result is that the entire net is divided into one metre sections and in each section there is looseness of webbing. The advantage of this net over that of an ordinary gill net is, while a fish gets entangled in one section the entire net does not get entangled as happens too often in ordinary gill nets. Even if one section is entangled, other sections are free and their catch ability does not get affected.

Trammel Nets

These nets consist of two or three walls of webbing hung together from a single float line and lead line. One wall of net is made of large mesh and hung tightly over the lines and other wall is of finer mesh and hung loosely. Fish, when caught, push the bag of fine meshed net through one of the larger meshes of the second webbing, thus forming a pocket, from which they are unable to escape, Trammel nets are costlier than ordinary gill nets and require more careful hanlings and their performance was net better compared to the ordinary gill nets in Mettur Reservoir.

Monofilament Nets

Nylon monofilament nets were tried successfully in Metthur reservoir and this efficiency was 10 to 20 times that of the local nets. Their relative lightness and their easy handling proved ideal for fishing purposes. Due to this transparency their visibility in water is practically nil. The only disadvantage with them is the difficulty encountered in mending them. Slippage of knots was common which hampered mending operations.

Fish Attractions

Attractions of fish by chemicals and artificial lights is popular 4n Japan and Russia. In the Caspian sea, underwater lamps are used successfully for 'Kilki', a clupeid fish, and on an arrange about 5-6 tonnes are caught daily. The fish is attracted by artificial underwater lamps and are pumped by means of hose in the boat. In India experiments conducted with electric underwater lamps of different colours, coloured nets (Table 32) and intensity and kerosene gas lamps above the surface of water were not satisfactory. Perhaps the indigenous carps are not of a shoaling nature and even their individual response to these lamps was poor. Only minnows like *Xenontodon* sp. and other minor carps were attracted by these artificial lamps.

More intensive study of the behaviour of fish for artificial lights is necessary, as that would help to do away with more cumbersome methods of fishing.

Table 32: Effect of Different Colour Nets on Fish Catch

Colour of Nets	*Catch kg./100 m²*
White	38.9
Blue	55.1
Green	66.9
Orange	83.4
Yellow	85.5

Source: Sugunan V. (1995).

Electrical Fishing

Electrical fishing will be useful in areas where the hazards of submerged obstacles are great and net fishing is impossible. Experiments conducted at Mettur and Krishnaraj Sagar reservoirs revealed that under low voltage large sized carps and catfishes get benumbed and come up to the surface of water and recover when the electrodes are removed. In reservoir proper, when the experiments were conducted poor results were obtained. In the open areas of the reservoir, the effect of electrical field was limited and consequently though the fish gil stupefied by the electric current, they swim away to places of safety where the field of electricity is not felt. These prove to be some of the handicaps for the successful application of the method in reservoir fishing. However, if the concentration of fish is gauged by use of eco-sounders, the efficiency of this method may increase.

Fish Detection

The navigational and detection instruments help to identify the concentration of fish and bottom types and measure the abundance of fish and detect submerged obstacles. For fish detection, ecosounders are used which are simple to operate. They were successfully tried in Metthur and Krishnaraj Sagar reservoirs and submerged obstacles could be easily located and nets were operated without any damage.

Chapter 23
A Ward on Reservoir Life

Apart from high silting, Srinjoy, Chawdhury in his recent article 200 dams on brink of collapse has high lighted certain facts on Indian reservoirs which is quite alarming (*The Statesman News Service,* 2004).

1. Of the country's 4050 dams nearly 350 are between 50 and 100 years old and 188 have recently been identified as distressed and in need of urgent repair.
2. 188 dams have different kind of weakness including leakage leading to seepage and in need of concrete grouting or non serious problems.
3. In some cases the face of the spillway show damages with the concrete shearing off. In some cases a flow in the design has been detected, especially after the dam was given additional capacity.
4. The damaged dams spread in the country are:

State	*Number of dams*	*Name of Dams*
West Bengal	2	Kangsabati–Kumari and the Massanjore Projects
Bihar	8	Gandak and Kosi, Jharkhand–till recently a part of Bihar has–5.
Orissa	38	Kumbhokalo, Damsel, Talkhol, Banksal, Ranjali, Talsara and Aradei
Uttar Pradesh	1	Rihand
Andhra Pradesh	9	–
Gujarat	18	–
Karnataka	29	–
Kerala	31	–
Madhya Pradesh	6	–
Maharashtra	13	–
Tamil Nadu	12	–

5. Eastern India has not only a large number distressed dams, but most of the country's flood prone areas. About 7.56 million hectares are affected every year on an average, a little less than half of it being cropped land,
6. Some damage is inevitable as many dams were built during British times. Some were ready by 1900 and between 1901 and 1950-281 were built.
7. Many huge projects happened in the fifties and sixties like the Bhakra Nangal and Damodar valley. Since 2001 only 4 have been readied,
8. Earlier the World Bank had helped to repair 49 dams, unfortunately no Central schemes to assist the states in repair of the dams. As a result the old dams may become problems to the country than help in future.

Chapter 24
Environmental Changes, Stressed Environments and Biotic Life

In reservoirs large volume of water is allowed to be stored up to a particular level during monsoon and depending upon the requirement (irrigation, power generation and water supply to industries and cities) the water is released. The littoral and sub-littoral zone gradually gets exposed and fail to contribute rich water quality appreciably. Although the river feed the reservoir round the year, its supply is gradually reduced till it reaches its minimum level around June/July until monsoon sets in. During late monsoon or draught period the reservoir water indicates different picture. Under such stressed ecological condition biotic communities are to grow and reproduce to help for secondary and tartiary production. During July and August when monsoon breaks the character of water (biotic and abiotic) gets diluted and change of ecosystem interferes the survival of saprobes. Even under such unstable condition macrobenthic invertebrates have to grow and their concentration is normally restricted to littoral and sublittoral zones around 10 m. depths. In case of smaller reservoirs such problems are not so acute on account of low fluctuation of water level.

Fry/Fingerlings do not find such ecosystem to be congenial for their survival and growth and they try to escape or gradually get extinct. Nagarjuna Sagar, Krishnaraj Sagar, Mettur Dam, Sarda Sagar, Nizam Sagar, and Amaravathy are some of the reservoirs where such problem exists.

Pollution is another important problem which considerably affects heating biomass production. Like rivers, riservoirs are also found to be affected by the complex industrial and municipal wastes. Ecodegradation of reservoir water from thermal power plants located near the reservoir and pollution load from various complex waste water either directly discharged or carried to the reservoir along with the stream water joining the reservoir. Broadly these are from chemicals (heavy metal like zinc, copper, chromium. Cadmium, lead, mercury etc. and pesticides) and thermal power plants/coal washeries disposing Coal washings. Survey made so far reveal that Gitalsud, Gandhi Sagar, Tungabhadra, Rihand, Bhavani Sagar, Hussain Sagar, Hirakud, Sandynulla, Panchet and Bhairomangola are some of the reservoirs facing such problems by one or the other. Heavy mortality of carp and catfish fingerlings, absence of fish food organisms, planklon and bottom biota have been recorded. These are discussed elsewhere.

Wide distribution of pesticide residues in water sediment, planklon and fish are reported. The residues of different isomers of BHC and DDT and its metabolites (DDE, DDD) were found in fish and fish food organisms in appreciable concentrations. The continuous influx of toxic effluents have adversely affected the food chain of Indian major carps resulting in dominance of resistance species. Such pollution problem affect the growth rate and fecundity, high parasitic infection and gut content containing mostly insect, insect larvae, worms etc. Feeding activity is reduced considerably. With the growing of industries which are likely to come up at a rapid pace in future such problems may increase considerably and as a result fish survival at all stages, growth and maturity may lead to a great problem to solve. Hence along with other study such problems may be dealt with seriously at the source by proper treatment of wastes.

Chapter 25

Dredging of River and Reservoir: A Measure of Improved Ecology for Fish Culture

Since reservoirs/dams have their direct link with rivers, maintenance is equally important. Flood in rivers and reservoirs due to dam failure (Bera, 1984) is a common phenomenon and it occurs in most of the states almost every year depending upon rainfall. Intensity of flood is dependant to rate of silting which increases year to year.

Dhruva Narayan and Rambabu (1983) reported that a total of 5334 million tones of soil is being annually eroded. The country's river carry an approximately quality of 2052 million tones, nearly 480 million tones are being deposited in various reservoirs nd 1572 million tones are washed into the sea. The chief minister of Assam observed worst floods in 50 years in the state (Chatterjee, 2003).

Even Dal Lake of Kashmir has reduced alarmingly compared to that of 50 years back (reduced to 12 sq. km in 1962, Kabita Suri 14.2.2007) The depth stands at 1.28 metre only at an average compared to 45 ft, 50 years back, stated by urban Development minister (Kabita Suri, 2006). Similar problems are also common both in North and South Indian reservoirs (Tables 6, 7).

According to the Chief Minister of Assam, dredging the rivers are more practical solution than a gargantium effort of coasting Rs. 5,00,000 crorse for river linking. Dredging may be expensive, but it is far cheaper than river linking.

With the growing population, every state has its own developmental problems and for which they require enough to water resources. The high flood water in most

Figure 13(a-l): Distribution of Reservoirs in Some States (Source: V.V. Sugunan, 1995, *Reprinted with Author Permission*)

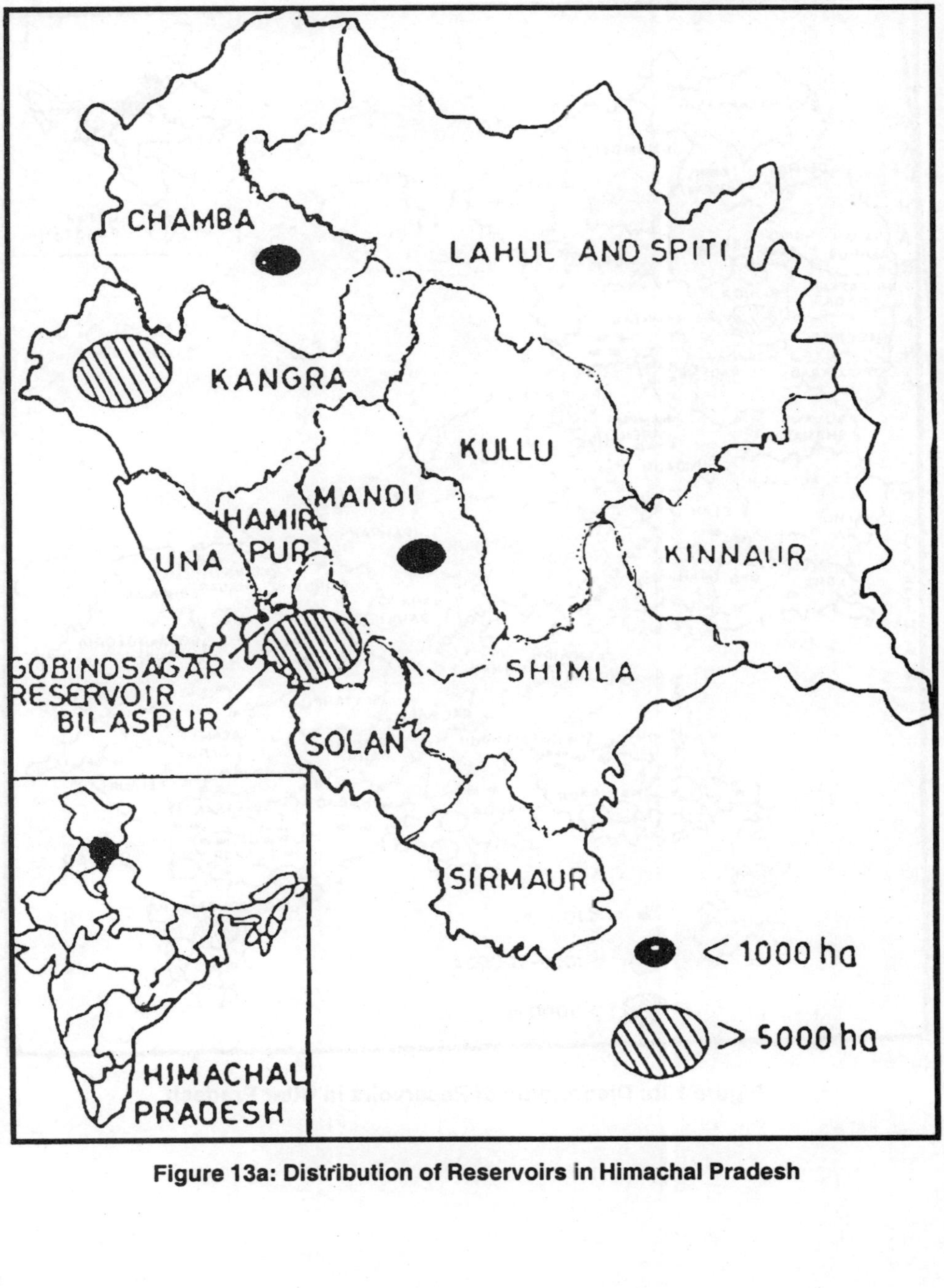

Figure 13a: Distribution of Reservoirs in Himachal Pradesh

Figure 13b: Distribution of Reservoirs in Uttar Pradesh

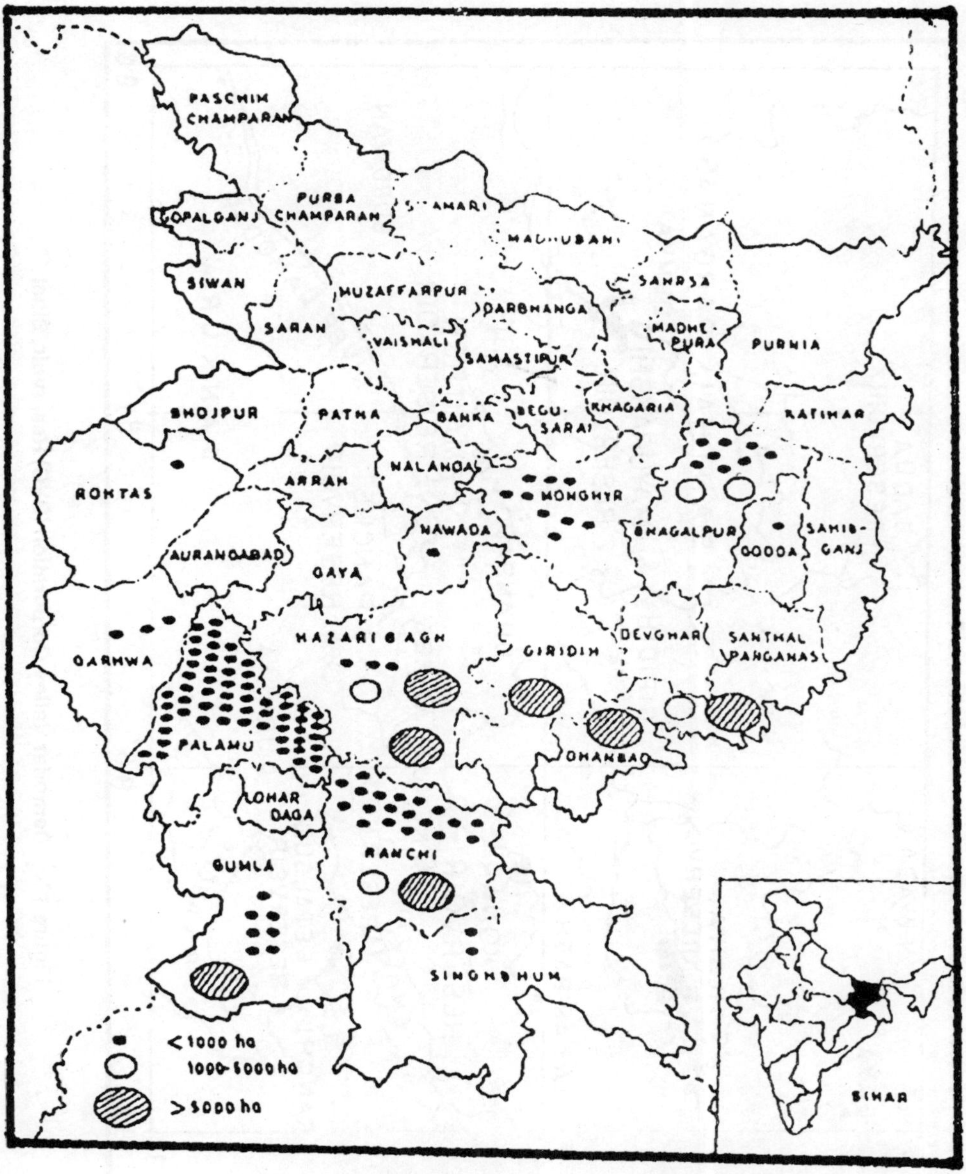

Figure 13c$_1$: Distribution of Reservoirs in Bihar

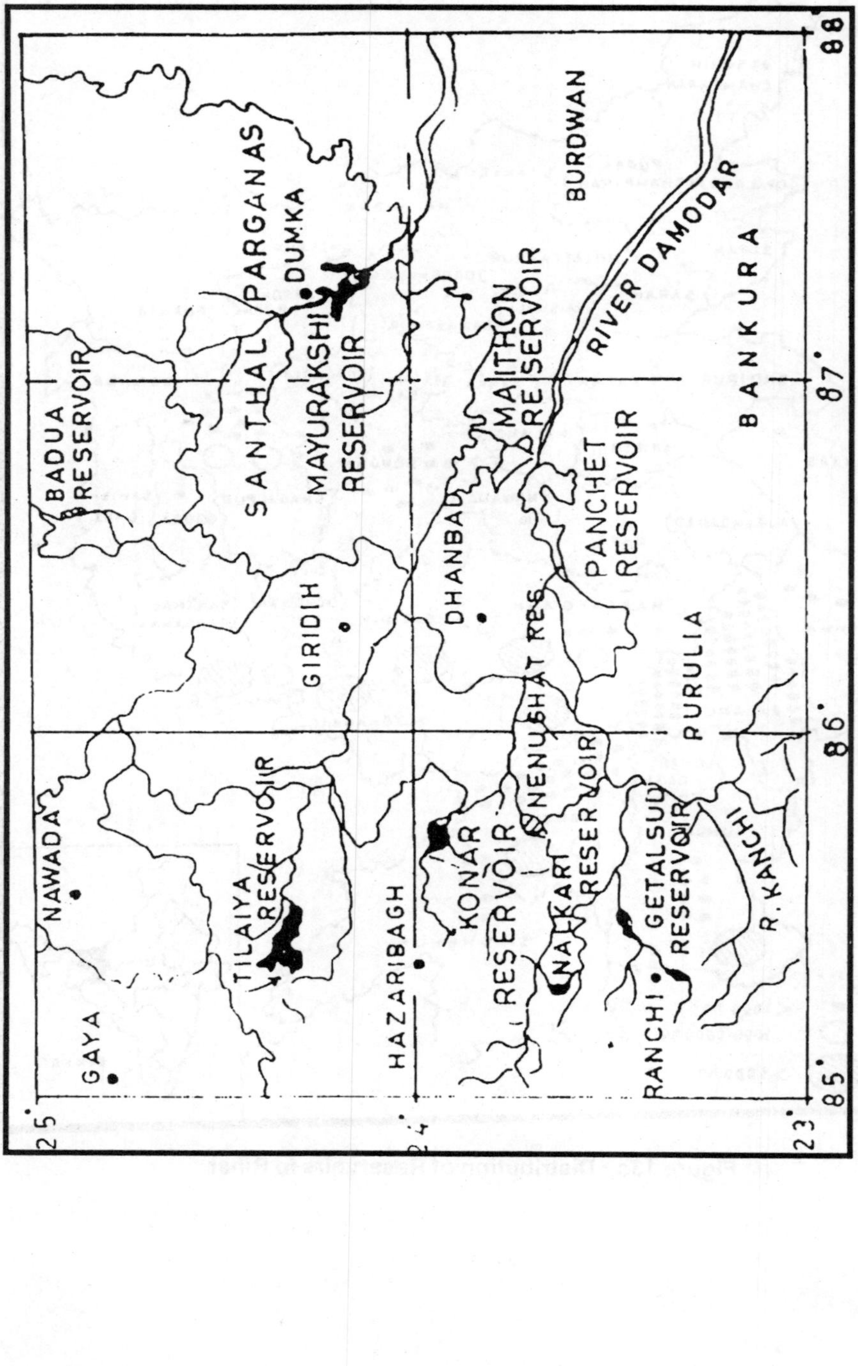

Figure 13c$_2$: Damodar Valley Corporation (DVC) Reservoir, Bihar

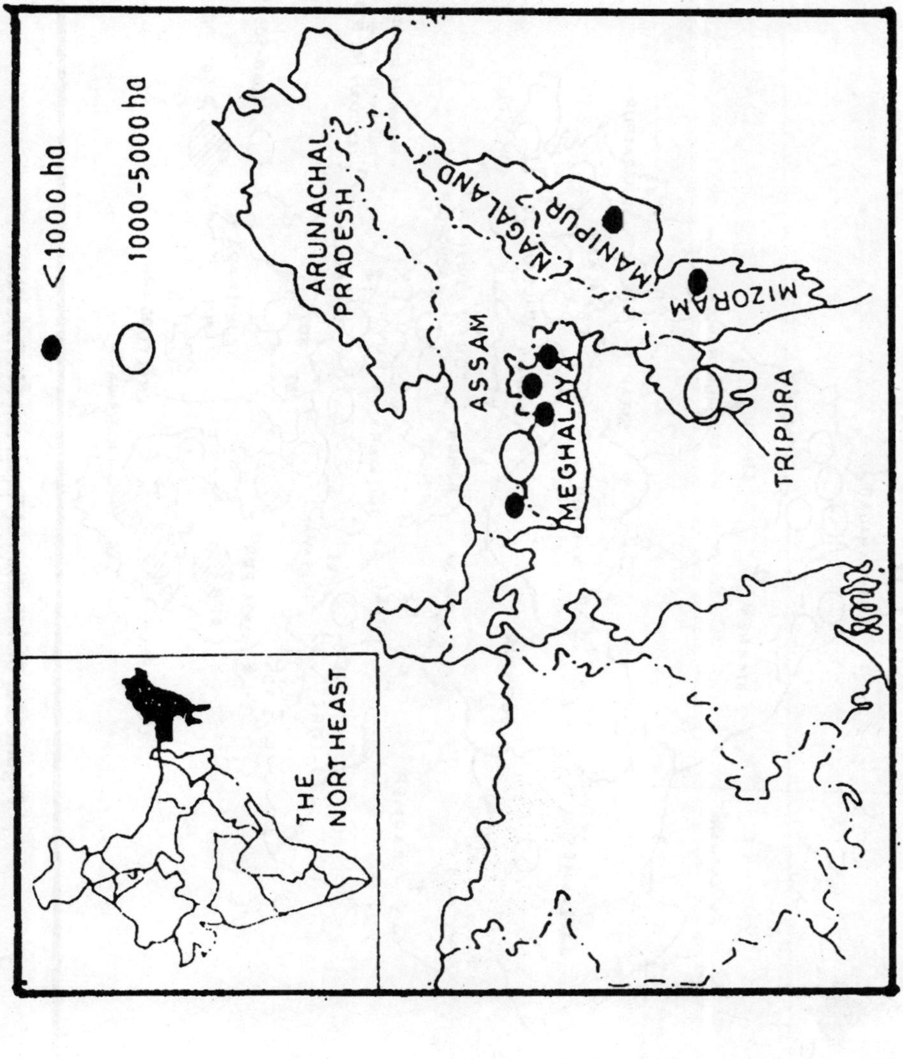

Figure 13d: Reservoirs of the North Eastern States

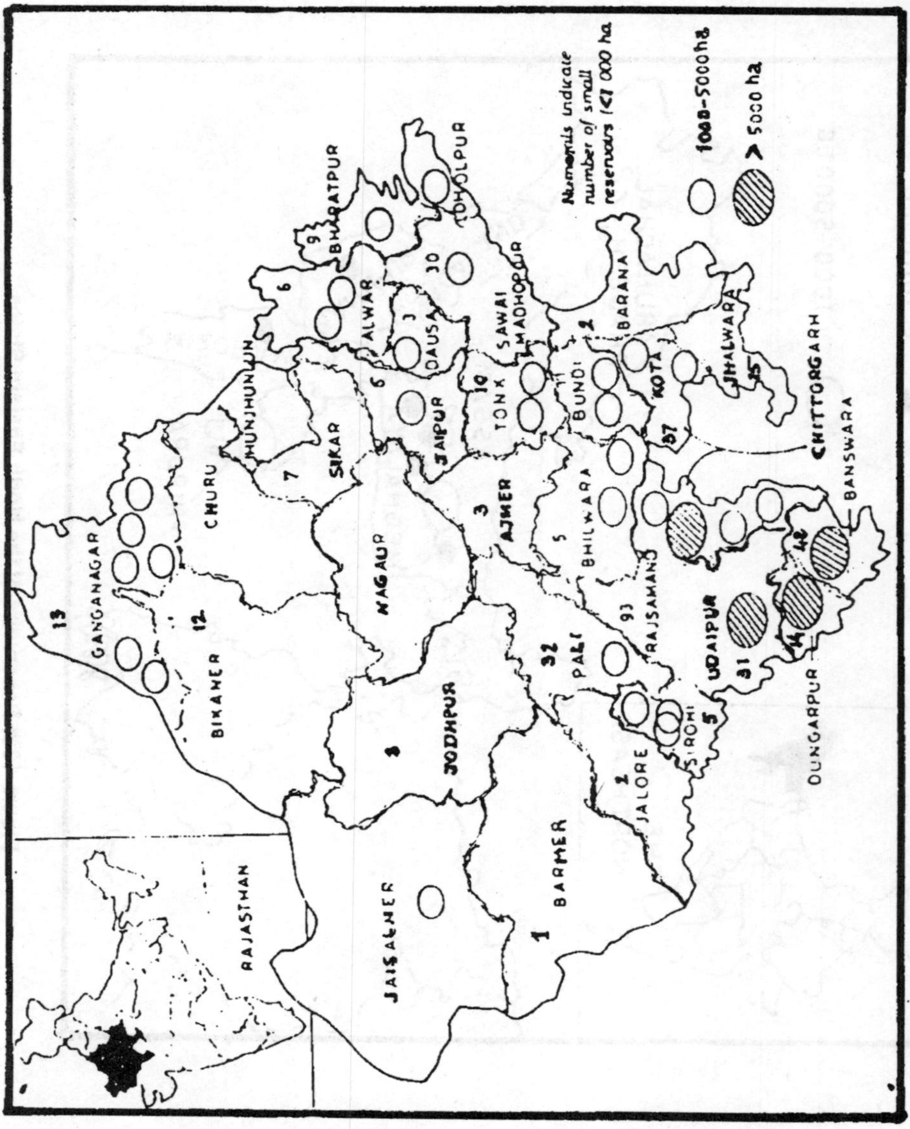

Figure 13e: Distribution of Reservoirs in Rajasthan

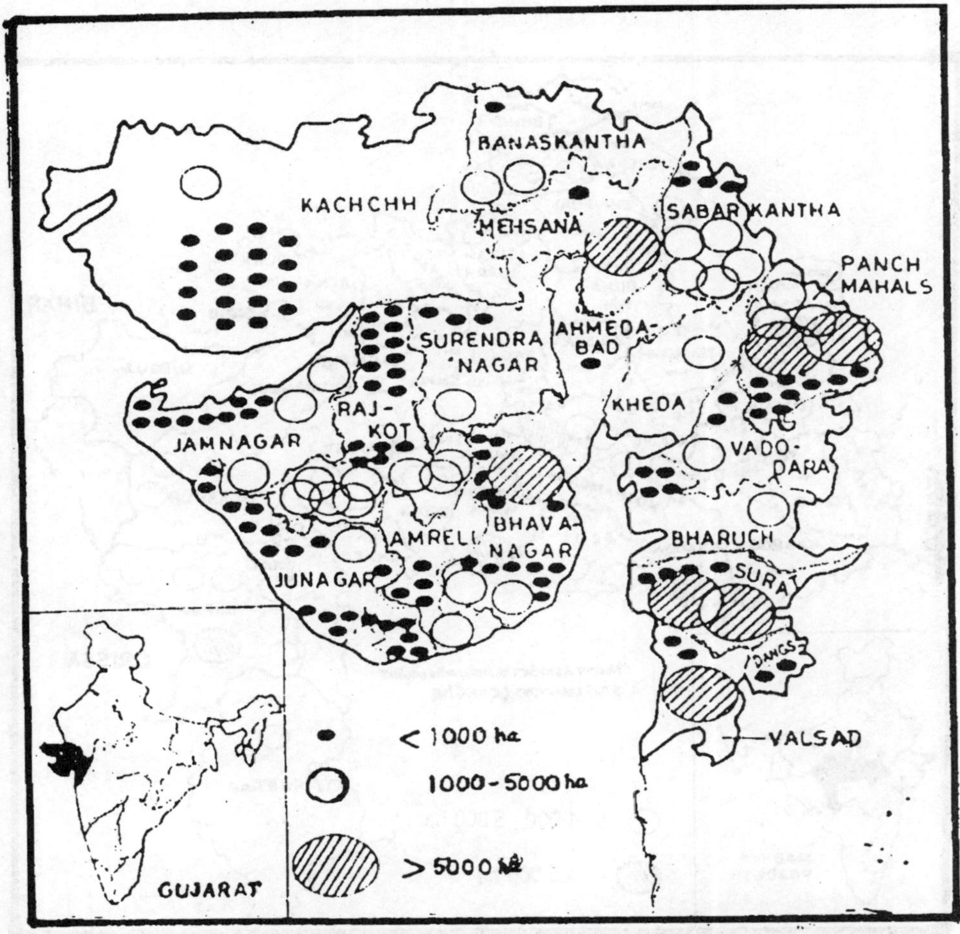

Figure 13f: Distribution of Reservoirs in Gujarat

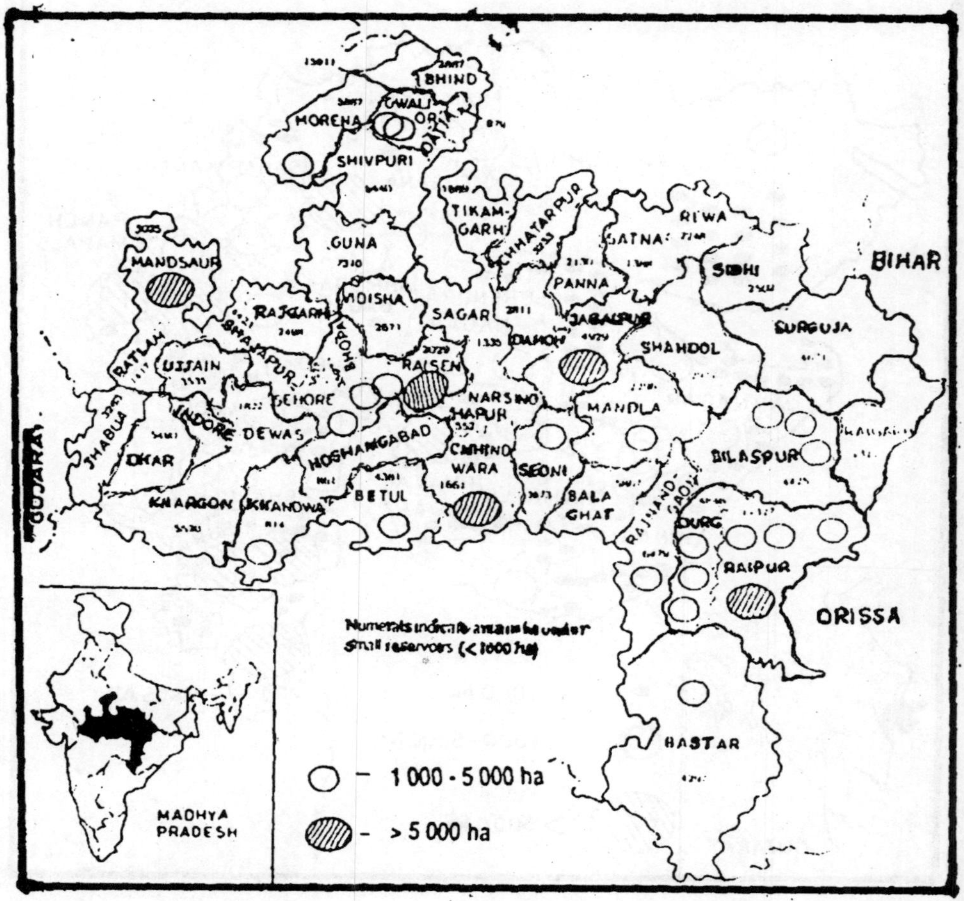

Figure 13g: Distribution of Reservoirs in Madhya Pradesh

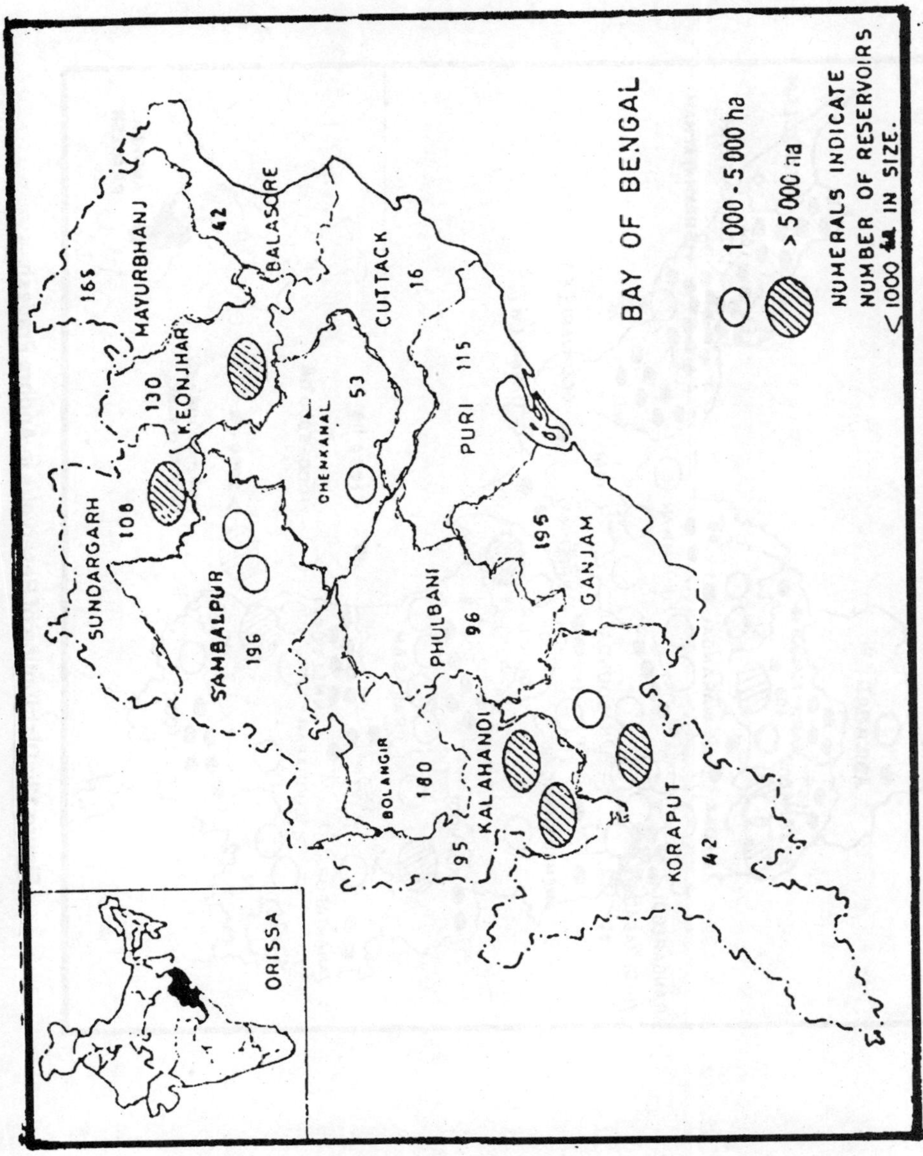

Figure 13h: Distribution of Reservoirs in Orissa

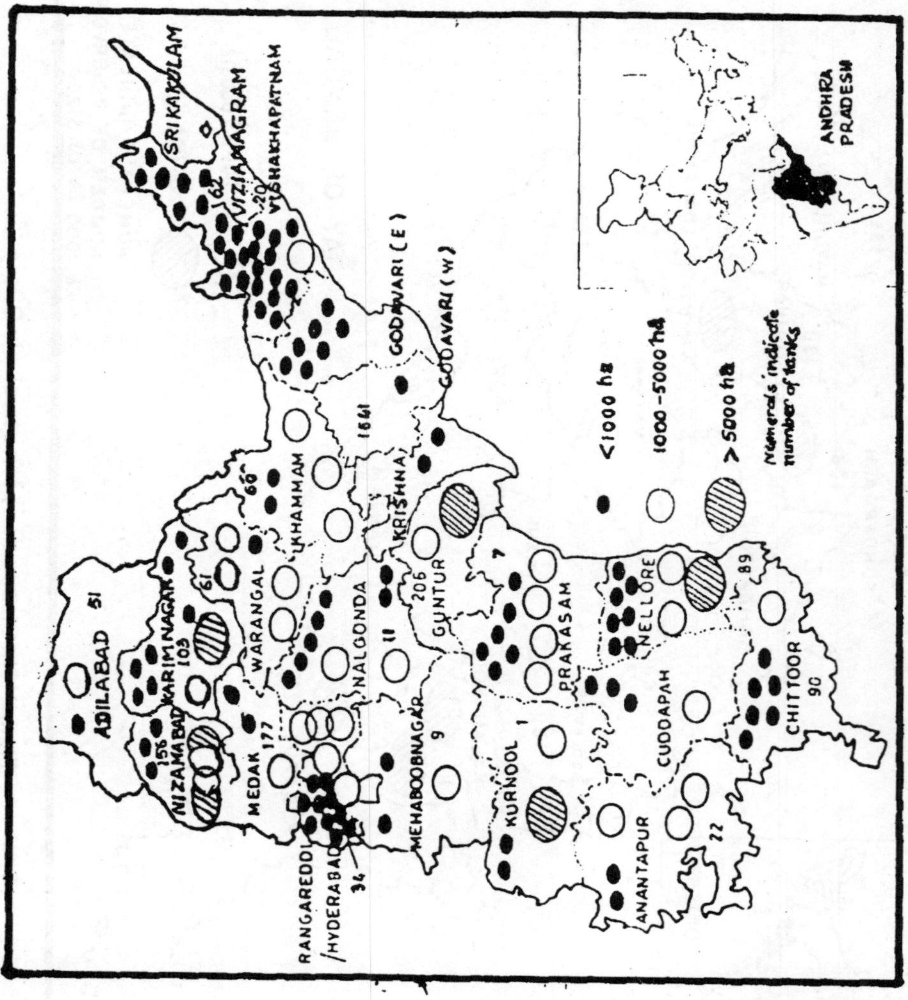

Figure 13i: Distribution of Reservoirs in Andhra Pradesh

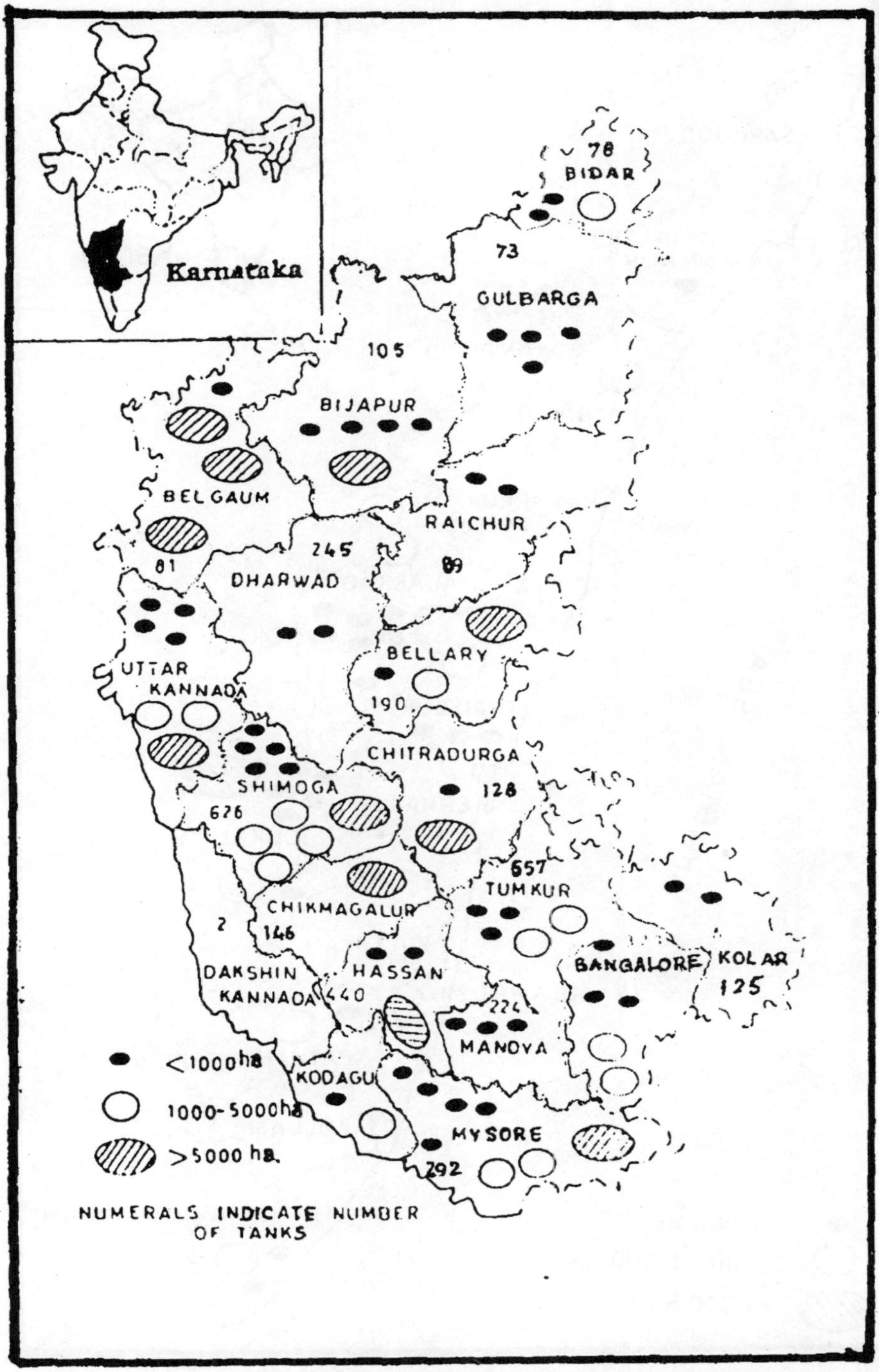

Figure 13j: Distribution of Reservoirs and Tanks in Karnataka

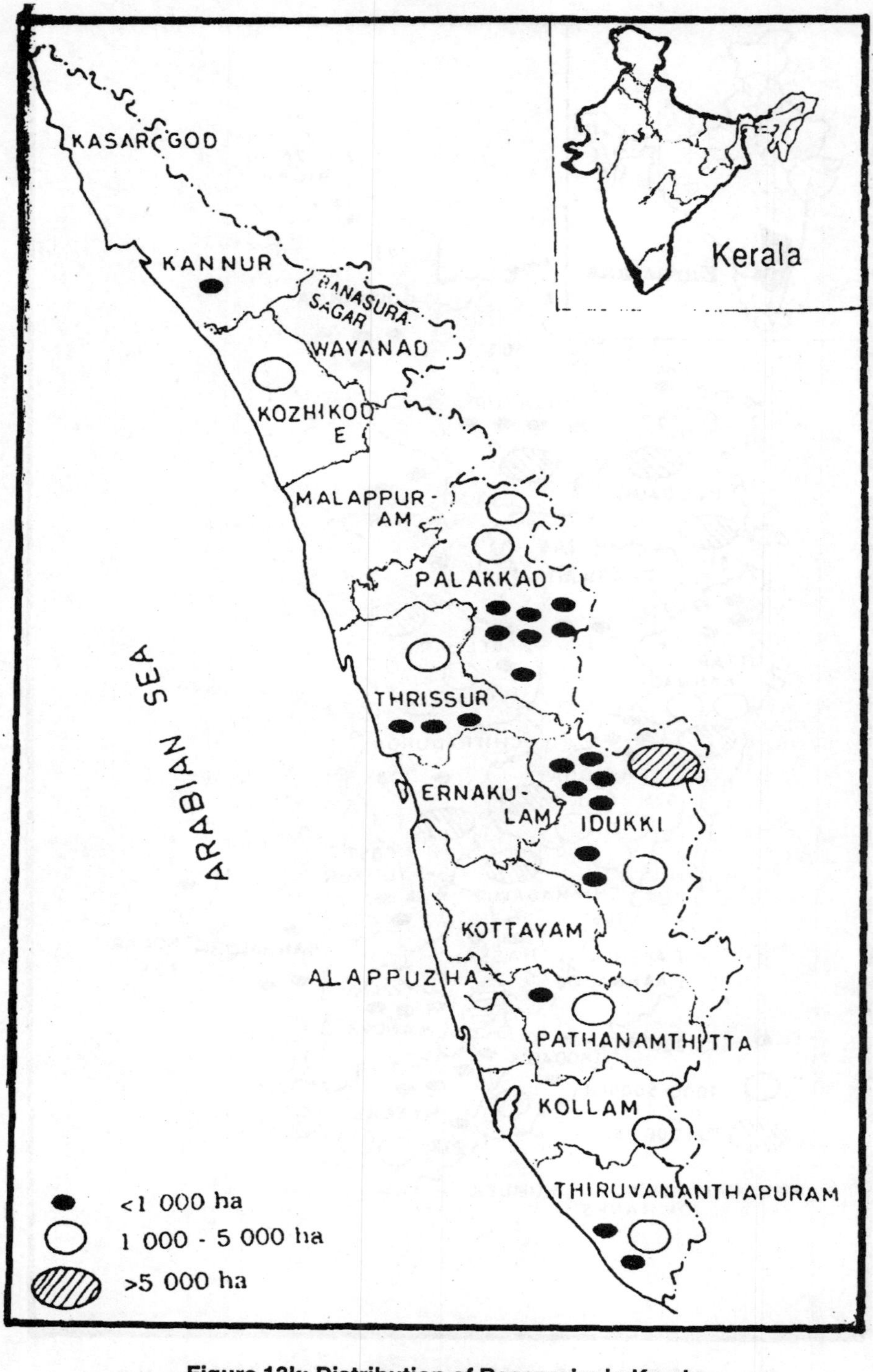

Figure 13k: Distribution of Reservoirs in Kerala

Figure 13I: Distribution of Reservoirs and Tanks in Tamil Nadu

of the states cause problem to the people and for which recurring expenditure (as relief measure) are incurred as shown below:

Table 33: Funds Allotted in Some States During 1993–1994 and 1994–1995 (in crore)

State	*1993-1994*	*1994-1995*
Punjab	11.24	15.7
Delhi	10	12
Haryana	8.9	9.1
Uttar Pradesh	9	8
Bihar	27	44.7
West Bengal	15	36

Source: Ministry of Water Resources (*The Statesman*, 14-4-1995).

If such flood water is controlled, the huge volume of water can be utilized during off season and simultaneously the recurring fund allotted can be save and utilized elsewhere. The appropriate solution of such floods perhaps is desilting of both rivers and reservoirs/dams including anciallary tanks (Figures 13a-l) to maintain adequate depth during lean period and afforestation to prevent silting.

In case desilting is maintained, the problem can be minimized appreciably. At the same time furtile soil can be used for various purposes (Agriculture, brick kiln manufacture etc.). Simultaneously, aquatic life both in reservoirs and rivers can grow well under improved ecological conditions.

Chapter 26

Hostile Atmosphere vis-à-vis Ecological Disorder

Atmospheric changes can alter the environmental conditions and as a result is like to have considerable impact on aquatic life. The weather is growing crazy, eratic and extreme and is becoming unpredictable. Such atmospheric changes are likely to alter the environmental conditions by way of high rainfall area changed into drought area, occurrence of flood in drought area, continuous rainfall for a few days, shifting of wet season, gradual fall of rains year after year, high atmospheric temperature (global warming), alarmingly low rainfall causing drought etc. are some of the aspects observed. These are discussed below:

1. Assam–Meghalaya, the eastern zone of India is known for heavy rainfall, causes flood in the adjoining vast areas, such worst flood in 50 years was recorded in July 2003, as a result 23, out of 25 districts were under attack of high water (Chatterjee, *The Statesman*, Kolkata, July 19, 2003.).

   The same area faced drought disaster in September 2006. A 30 per cent deficit in rainfall has caused a prolonged dry spell in 22 of Assam's 27 districts, for which Assam Government announced 50 crore special package (*The Statesman News Service*, Kolkata September 1, 2006).

   The following table indicates how rainfall is declining year after year in this area.

Assam		Meghalaya		Sub-division
		June		July
2001	...	–23 per cent	...	–29 per cent
2002	...	–21 per cent	...	–22 per cent
2003	...	+ 4 per cent	...	–27 per cent
2004	...	–8 per cent	...	+ 6 per cent
2005	...	–23 per cent	...	–28 per cent
2006	...	–40 per cent	...	–37 per cent

Above and below the normal rainfall.

Source: Premangshu Chaudhury and Debasis Chaudhury. In: Ananda Bazar Patrika, September 6, 2006.

2. Contrary to this is also recorded in Gujarat and Rajasthan all dry areas where rainfall is negligible/desert has flooded this year (2006) considerably due to continuous down pour–a rare example.
3. Besides these high flood was also observed in Bombay (Maharashtra) during 1959 and 1960, Kashmir, Himachal Pradesh, Punjab, Rajasthan, Gujarat, Madhya Pradesh all during 1960 due to continuous down pour for a few days. Continuous downpour associated with thunder storm and heavy snowfall has upset the normal life of Jammu and Kashmir. Snowfall which was 72.3 mm a couple of days before increased to 89.0 mm a record of 15 years. Even Himachal Pradesh, Delhi, Uttarakhand, Uttar Pradesh, Punjab and Haryana have become upset due to heavy storm resulting considerable fal in temperature (*Bartaman*, 14.3.2007).

Contrary to this is also on record. There were regular heat waves at Jammu which reached as high as 39°C, 6° above the normal highest temperature. At Sreenagar the highest temperature attained was 29°C while at Himachal and Uttarakhan the highest temperature level ranged between 24 and 25°C.

Unnatural rise in temperature in North Pakistan has affected Jammu and Kashmir, Uttarakhand, Himachal Pradesh by unusual rise in temperature, being adjoining area of North Pakistan. Even the hilly area of Haryana and Rajasthan conditions are still worse. At Jaipur (Rajasthan), Jaisalmer the temperature attained up to 42° and 45°C which means a further rise of 5°C from the normal upper level. Haryana topped the list when the temperature rose as high as 44°C and 7°C more than the normal temperature of the upper level (*Ananda Bazar Patrika*, 18.4.2007).

Continuous high atmospheric temperature (global warming) faced with its draught in 30 years in Karnataka. The South West monsoon has brought 42 per cent less rain fall this year resulting severe damage to Krishna Raj Sagar (Karnataka) which is never recorded earlier (Manoj Ghosh, *The Sunday Statesman*, Kolkata, 29 September, 2002).

In DVC area with temperature hovering in the regin of 44°C for the past four days in Burdwan (West Bengal) rsulted ground water to come down drastically below 1 m. mark in some places in Kalna, Katwa, Bhatar and Mongolkote blocks. Even in reservoir (Mithon reservoir) large areas got exposed, is also a rare case (*The Statesman News Service*, Kolkata, 3.6.2005).

Are all these the resultant effect of global warming which Prof. R. Brahmachari of Calcutta University, predicted in his valuable article Global warming could cause serious damage in India under "Danger signal-II" (*The Statesman*, Kolkata 17.1.2006).

Under such unusual variation in the atmosphere which results abnormal change in the ecosystem observed in the form-late monsoon, continuous down pour (even in drought area) and flooding, high silt precipitation probable change in water quality, over flowing of reservoirs, drought-resulting fall in the water level etc. are likely to alter survival, growth, maturity and reproduction of the established aquatic life of reservoirs and lakes etc. are some of the major problems that may result.

Chapte 27

Conservation and Management

Conservation and management is thus an important aspect which requires adequate attention to enhance the productive potentiality of reservoirs. The following are some important aspects.

1. Monitoring of catch and fishing effort is a must. For this the free fishing may be replaced by a system of licensing the nets and boats to enable regulation of fishing effort,
2. The destruction of large scale of fishing migration of brood fish should be stopped«
3. In case any species of fish do not come up in any particular reservoir it may not be stocked.
4. Careful attention may be given to check or control the miscellaneous uneconomical trash fishes like catfish or any other predatory fishes to encourage growth of fish which are commercially important. But if in any particular state or zone has any preference for any particular species like catfish, murrels etc. they may be cultured. These fishes on the other hand may consume the undesirable weed fishes as food which is largely present in the reservoir.
5. Regular monitoring of the catches may be made and take a note of different species to take steps for proper stocking, which means whether different ecological niches of the reservoir are being properly covered by species of fish stocked. Whether there is any recruitment by natural breeding or any recruitment success has to be ascertained. These points are important while stocking the reservoir scientifically. Growth rate determination of individual species is essential.
6. Experimental fishing may be conducted with different nets having mesh bar ranging 25-180 mm commonly used in reservoirs to identify areas of

fishing dispersal in time and space. Uniformity of all classes of fishing gears in all the reservoirs should be encouraged for their standardization. This will help to ascertain the fish production and compare the yield rate. The National yield rate per ha per year can thus be determined and any other effort to improve the production can be thought of. This is an important aspect.

7. Occasional monitoring of water quality (biotic) is necessary to ascertain their diversity and abundance of periphyton and other benthic forms that favour of stocking of Indian major carps or any other commercial important species.
8. Industries and cities discharging complex wastes (metallic and Pesticide) into the river joining the reservoir or directly into the reservoir should be treated adequately before releasing them into the water. In such cases monitoring of water quality (biotic and abiotic) should be done to ascertain that the biomass including fish is free from bioaccumulation from metals and pesticides. In this connection the author wants to provide additional information for controlling pollution. Although non-polluting energy resources are being developed and technologies being diverted to minimise impact of pollutants they are not enough to cope with the situation. Hence, besides the existing traditional system of treatment of such complex waste water, there is a tree "strychnos Pastatorm", locally known as "Chilla Ginjalu" which largely grows in the forest of Andhra Pradesh, has excellent purifying capacity for clearing the effluents discharged from various factories as well as from Atomic Power Plants, may be tried. A large scale plantation of this tree at the appropriate place may be encouraged both by the Government and other welfare agencies. The details of the "wonder tree" is also enclosed in appendix.
9. Certain section of rivers and reservoirs if possible can be kept closed for fishing for a certain period or round the year where fish congregate for breeding or fry and fingerlings are found in large numbers (Sanctuary).
10. In all large reservoirs fishing may be closed from June/July to end of September so that fishes are not hampered during their spawning migrations and allowed to breed. Large scale destruction of fish takes place when they migrate upstream of Tilaiya, Hirakud, Pelwa and Kotwal reservoirs for breeding. It is reported that at Tilaiya over $1^1/_2$ tons/day of fishes were caught by local population in the upper reaches of the reservoir and in Barakar river.
11. Poaching has to be controlled as it affects the net production of a reservoir. In case of a smaller reservoir poaching is easy and the entire production can be removed in a few nights.
12. Waste weirs and sluice gates of canals of reservoirs cause considerable loss to the fishery during monsoon when reservoir water touches the highest level (Palwa dam loses about 36 kg/day of fish). This has to be checked by installing trush rocks or other suitable methods.

13. In most of the villages in the vicinity of the reservoirs or on their periphery there is hardly any fishermen population. It has been observed that fishermen parties from U.P. migrate to Rajasthan for fishing. At least 400 fishermen are required for exploitation of the reservoir–a scheme for the rehabilitation and organisation of co-operative societies are required to be set up.
14. At the time of impoundment houses, forest tracts, rocks, boulders etc. get submerged which create lot of difficulties during net operation, Hence for effective operation of net such obstructions are to be cleared at the time of planning.
15. Productive zones (especially large bays) are to be protected as their production is significantly higher than the entire reservoir production. This is shown by Dubey (1969) during the course of his studies.
16. Deforestation has to be controlled to check high silt-erosion by adopting proper afforestation scheme. It has been seen that annual silt precipitation in reservoirs has shown many times more than assumed silt precipitation. The high silt precipitation changes the morpho-ecological condition of reservoirs and at a later stage they may loose their character for which it was constructed. Water holding capacity is lost and with normal rainfall they reach the danger level when release of water becomes essential. By this process lot of fish escape. Further silt precipitation affect the breeding grounds adversely and affect the biotic communities considered for fish.
17. Such alluvial silt may be shifted elsewhere by occasional dredging, for filling the low lands, use as furtail soil in local agriculture fields and also for manufacture of bricks, as are normally done in the Gangetic West Bengal (Hooghly estuay).

Chapter 28

Suggestions for Future Investigations

1. Priority in the latitude induced variation in the availability of Indian Reservoirs should be given, as Indian reservoirs are exposed to wide range of climates (temperate Himalayan) and extreme tropical in peninsular area. This may help in determining the status of reservoir energy region-wise. Data obtained from this can be grouped zone-wise.
2. Detailed investigation on the seasonal dynamics (Physical, chemical and biological) of the reservoir biota are essential factors and efforts may be made to standerdise such characters among reservoirs keeping in view of the morpho-ecological characters of reservoirs in different regions.
3. From the broad analysis of various factors responsible for the absence of some commercial important species in reservoirs it appears that the factors are mainly:
   (*a*) Changes of hydrological condition of the reservoir.
   (*b*) Restricted facundity and
   (*c*) Proper size of the stocked fish etc.

   All these factors are to be thoroughly studied in order to reveal more detailed information as these factors are directly related to fish production. Further to what an extent these factors affect the growth rate of different commercial important species and in case any positive effect, may mean changes in the physiological condition of fish for which haematological, serological, pathological and Cytological status of fish may be determined to find out the proper cause of their elimination reproduction and growth from a particular reservoir.

4. Since Indian reservoirs are mostly subjected to higher fluctuations in the water level, it is likely to have some impact on the healthy growth and maturity of fishes. Such facts may reveal valuable information on the extent of loss in fish production.
5. In view of the fact that plankton, fish and other benthos are likely to be affected by metals and also by pesticides through their bioaccumulation, studies on haematological, serological, histopathological and cytochemical tests are essential to find out any changes in their physiological aspects which may affect the healthy growth, survival, maturation and reproduction. A systematic monitoring programme is necessary for a proper assessment of man's impact on the reservoir water. Effluents need be treated before their release in the receiving water.
6. Pen and cage culture practices in proper areas (mostly in bays) may be encouraged to raise the fry/fingerlings to a proper size before releasing them into the reservoir. Fry/Fingerlings raised in hatcheries are supposed to be delicate and may not find the ecological condition of the reservoir to be congenial if directly released.
7. From different investigations carried out both in reservoirs located in tropical and subtrophical regions, it is observed that Rangoon-fishermen Party using only Rangoon nets (average specification: length–100 ft; Depth–15 ft; mesh bar 4 "-7.5" they also operate 2" bar gill nets) are found to be very effective and hence a standarisation of such nets using different meshes may be made that may give effective catches in all reservoirs. Any other gear *e.g.* new techniques of fishing suggested by FAO experts Gulbadamov (1962) and Znamensky (1967) can be tried to find out their effectiveness in catching fish and if found suitable can also be standerdised as suggested in case of Rangoon nets.
8. After proper stocking whether shore development index of reservoirs of different grade and volume development index (high and low, indicating concave or convex basin) have any co-relation or not with fish production may be studied thoroughly especially in medium and large reservoirs. Such studies may reveal valuable information.
9. Stocking, the most vital aspect should be done appropriately after knowing the existing fishery status of the reservoir as suggested by Jhingran and Natarajan (1969) during their studies of DVC reservoirs. Besides, how to enhance the growth of nursery rared fry/fingerlings to a maximum healthy growing fish in the existing reservoir ecosystem may be studied.
10. It is noticed that some economically important species after their transplantation in reservoirs fail to survive where as in some reservoirs they can establish well. Such problems may be investigated thoroughly.
11. Expression of production (yield/ha) should be uniformly stated. AS suggested by IMMA (1983) *i.e.*, production per/ha has to be taken as 60 per cent of the FRL because of high fluctuality water level or average area between FRL and DSL as suggested by NCA.

Chapter 29

Recommendations of the National Commission on Agriculture, Fisheries in their Report 1976

Reservoir Fisheries

1. Improvement in capture fisheries will be possibly only through refinement in fishing techniques which should be solved by Craft and Gear Research section of the CIFT (ICAR).
2. Preimpoundment surveys of rivers should help formulate downstream ichthyofauna. The proposals should be jointly developed by fisheries and irrigation organizations in the states for inclusion in the river valley projects and the required works should then be got done prior to improvement.
3. Reservoirs and tanks built for domestic water supplies should also be utilized for fish rearing but with adequate health safe guard.
4. To keep the continuity of stock of major carps which take long periods for acclematization for natural breeding, it should be ensured that the adequate level of water is maintained in the dead storages of medium and major reservoirs.
5. Fishing in the reservoirs should be rationalized by systematic removal of trash fish and observing close seasons to stop indiscriminate killing of brood fish, particularly of major corps.
6. To maintain the productivity of reservoirs, the possibility of cultivating suitable legumes in the foreshore areas exposed during summer months

should be explored. The top cuttings of these crops would be useful as green fodder.

7. The fish seed supplies of major carps have to be increased several fold, by all methods of production for undertaking increased stocking densities to obtain optimal yield from all culturable waters.
8. Necessary conservation measures should be undertaken to protect breeding grounds of major carps. Riverine stretches which harbour large scale concentration of seed fish should be kept only for fish seed collection.
9. Intensify seed farms near reservoirs and other areas where parennial water facilities exist.

Appendix

Wonder Tree from Andhra Controls Pollution

A substance isolated from the seeds of "Chilla Ginjalu", a tree growing in the forests of Andhra Pradesh, may provide the answer for controlling global pollution, scientists say. The substance also binds 18 metals including uranium and thorium. Besides, it helps in cleaning effluents from factories as well as atomic power plants.

For centuries, tribes in Andhra Pradesh have used the powdered seeds of the "Chilla Ginjalu" (which also grows in Myanmar and Thailand) to purify drinking water. The powder had the ability to bind itself with impurities and remove them–a phenomenon called "flocculation".

By adding modern biology to this tribal secret, scientists of the Andhra Pradesh Grijan (Tribal) Cooperative Corporation, have now isolated from the seed a "bio-flocculant".

The tree (Strychnos Postatorm) that was hitherto unknown outside the region, is now up for international exposure with industries world wide and the International Atomic Energy Agency in Vienna showing interest in the product,

"It is non-toxic and bio-degradable", says Mr. Durgaprasad, one of the scientists behind the operation. Apart from uranium and thorium, the substance easily binds with toxic mercury and cadmium that are usually present in factory effluents.

It also tags on to metals such as silver, gold, copper and nickel, thereby providing mining companies an entirely new method of ore recovery, according to Mr. Prasad. He said the metal binding nature was because of proteins present in the bio-flocculant.

According to the State Chief Minister, Mr. Vijaya Bhaskar Reddy, the product will be soon commercialized and profits ploughed back for tribal development, Andhra Pradesh has a tribal population of 4.2 million people and over 44 per cent of them live in forests.

According to an American scientist, moves were on for collaboration aimed at further purifying the proteins in the bio-flocculant and improving its metal-binding ability Dr. C.E. Furlong of the American University at Seattle says that the protein isolated from the Indian tree was unique because some of the substances developed in the United States could at best bind with one or two metals, and not 18,

Dr. Furlong added his university plans to use genetic engineering methods to synthesize the proteins found in the seeds of the Indian tree. Dr. K. Mahadev Rao of the IAEA says substances isolated by GCC scientists had great prdmise in removing traces of uranium and other metals in reactor effluents.

He said its ability to remove cobalt was particularly useful to nuclear engineers as the steel used in nuclear industry must be cobalt-free. Dr. Rao said the substance can also bind with "actinides", a group of metals in nuclear dumps which emit radiation for a long time. These can be selectively removed using the GCC's bio-flocculant.

According to the GCC managing director, Mr. T. Vijay Kumar, the research on "nut-cleaning" was started three years ago by himself and four of his colleagues in a car shed. There was now a stock of 2,000 tons of seed which can produce the substance on a semi-commercial scale.

The local scientists have also developed the technology for immobilizing the bio-flocculant on a suitable polymer, which means that it can be re-cyclede–DPA

The Statesman, Tuesday October 26, 1993

References

Achuthan Nair 1969. On the unusual migratory habit of the fishes of Malampuzha Reservoir. *Proceedings of the Seminar on the ecology and Fisheries of Fresh-water Reservoir*, November 27-29. ICAR Krishi Bhavan, New Delhi-11001, at CIFRI, Barrackpore, W.B. India..

Agarwal, M. and H.D. Kumar, 1978. Physico-chemical and Physiological assessment of two mercury polluted effluents. *Indian J. Environ. Hlth.* 20: 141-155.

Anonymous 1984. Water management: Goal, Policies, Objectives and implementation, *Procedures of the Ministry of Environment*. Ontario, Canada.

Arora, H. C., S. N. Chattopadhyaya and U. P. Sharma (1970). A probable occurrence of fish mortality in Renu Sagar, Renukoot due to Chlorine bearing wastes. *Environ. Hlth.* 12: 260-272.

Babu Rao, M. 1990. Effect of Entrophication on the Fish Biology of Fishes in Hussain Sagar. A case study. *National workshop on Reservoir Fisheries* 3-4 January.

Asian Fisheries Society, Indian Branch. Central Inland capture Fisheries Research Institute, Barrackpore, West Bengal..

Balon, E. K. 1975. Reproductive guide of fishes–A proposal and definition. *J. Fish. Res. Board Cen* 32: 821-864.

Bandopadhyay, M.K., S.K. Singh and B. Sarkar, 1991. Growth and Production of Ctenopharyngodon idella reard in plastic cages. *Prof. Nat. Symp. Freshwater Aquaculture* CIFA, Bhubaneswar, 35-37.

Banerjee, R.K. and Ashoke Mukherjee, 1990. Man-made Ecological detesioration in Panchet Reservoir, Bihar (Abstract) *National Workshop on Reservoir Fisheries* 3-4 January Asian Fisheries Society (Indian Branch) Central Inland Capture Fisheries Research Institute, Barrackpore, West Bengal.

Benson, N. G. 1968. Review of fishery studies on Missouri river main stream reservoirs.

Bera, S. 1984. *'NADIBIGVANER KATHA'* (The Science of Rivers) 10, Raja Ram Krishna Street, Calcutta–700 006.

Bhardwaj, S.P. and Ravender Singh, 1985. Environmental Preservation through soil and water conservation. *National seminar on Pollution Control and Environmental Management*, 16-18 March, NEERI, Nagpur, Vol.V. Environmental impact Assessment organized by NEERI, Nagpur and Indian Association for water Pollution Control, Nagpur. Co-sponsored by Department of Environmental, New Delhi.

Bhukaswan, T. 1978. The Development and Management of Freshwater resources.

Brusven, M.A. and K.V. Prather, 1974. Influence of stream sediments on distribution of macrobenthos. *J.Entomol. Soc.* 71: 25-32.

Final Report, 1971-81 IFRI Series 1 May 1981. All India Co-ordinated Research Project on Ecology and Fisheries of Freshwater Reservoirs Information Series 1, May 1981, Bhavani Sagar.

Final Report 1971-81 IFRI Series 3 March 1983. All India Co-ordinated Research Project on Ecology and Fisheries of Freshwater Reservoirs, Nagarjuna Sagar. Inland Fisheries Research Information series 3 March, 1983.

Final Report 1971-81 IFRI Series 4 March 1984. All India Co-ordinated Research Project on Ecology and Fisheries of Freshwater Reservoirs, GETALSUD. Inland Fisheries Research Information Series 4 March 1984.

Chacko, P. I. 1969. Fish and Fisheries of the Muthupet saliae Swamp. Tanjore District. *Proc. Indian Sci. Congr.* 36(3): 166.

Chandra, K., R.S. Panwar and D.N. Singh 1985. Possible Pollution Problems from wastes of Thermal Power Plants in India. A case study of Rihand Reservoir Mirzapur (U.P.) *Proceedings of the synaposium on Environmental pollution*. pp. 283-294. The Academy of Environmental Biology Publication, Muzaffarnagar, U.P.

Chatterjee, Shib Sankar, 2003. Wast floods in 50 years, in The Statesman, 19.7.2003.

Chowdhury, Srinjoy 2004. 200 dams on brick of collapse. In: *The Statesman News Service*, 26.9.2004.

Chopra, R.N.R.L. Randhawar and S. Ghosh 1949. *Poisonous Plants of India*. Vol. 1, ICAR, New Delhi.

Chutter, F. M. 1969. The effects of silt and send on the invertebrate fauna of streams and rivers. *Hydrobiologia*, 24: 57-76.

Cordone, A. J. and D. W. Kelley, 1961. The influences of inorganic sediment on the aquatic life in Streams. *Calif. Fish Game.*, 47: 189-228.

David, A. 1969. Preliminary Survey Report on the Hydrology and Fisheries of Nagarjuna Sagar, Andhra Pradesh. Survey Report No.5. *Cent. Inl. Fish Res. Inst.* Barrackpore.

David, A. and K.V. Rajagopal, 1969. Experimental Fishing and observations on Commercial Fishing methods. *Proceedings of the Seminar on the Ecology and Fisheries of Freshwater Reservoirs* November 27-29. Sponsored by ICAR, Krishi Bhavan, New Delhi–110001 at CIFRI, Barrackpore, West Bengal, India.

David, A., N.G.S., Rao and P. Ray, 1974. Tank Fishery Resources of Karnataka. *Bulletin Central Inland Fisheries Research Institute,* Barrackpore, p. 87.

Desai, V. R. and R.K. Singh, 1979. Occurrence of Thermocline in Rihand Reservoir of Uttar Pradesh and its depth criterion in relation to Reservoir level; *Jour Inland Fisheries Soc. India* Vol.II, No.2, Dec. (Published in 1980).

Devaraj, S.P.M.G. Prabhavathy and P.K. Goswami, 1990. Assessment of Optimum Fish yields from reservoirs in Tamil Nadu. *National Workshop on Reservoir Fisheries.* Abstract 3-4 January, 1990. Asian Fisheries society Indian Branch. Central Inland Capture Fisheries Research Institute, C/o. CIFRI, Barrackpore 743101, West Bengal.

Dhruva Narayan, V.V. and Rambabu, 1983. Estimation of soil erosion in India. *Journal of the Irrigation and Drainage Engineering* 109(4), M.419-434. In: Environmental Preservation through soil and water conservation–Bhardwaj, S.P. and Revender Singh.

Dubey, G.P. 1969. Development of Reservoir Fisheries in Madhya Pradesh. *Proceedings of the seminar on the ecology and Fisheries of Fresh water Reservoirs.* November, 27-29. Sponsored by ICAR, Krishi Bhavan, New Delhi–110001 at CIFRI, Barrackpore, West Bengal, India..

Dubey, G.P. 1969. Development of Fisheries in Reservoirs. (Memo).

Dubey, G.P. and S. N. Chatterjee 1977. Case study of Gandhi Sagar Reservoir, Madhya Pradesh, India. *Proc. IPFC,* 17(3): 232-246.

Fish, G.R. 1970. A limnological study of four lakes near Rotorun, N.Z.I. *Mar.* and *Freshwater Res.* 4(2): 165-94. In: Occurrence of Thermocline in Rihand Reservoir of U.P. and its depth criterian in relation to Reservoir level–V. R. Desai and R. K. Singh. *Jour. Inland Fish Soc.* (India) Vol.11, No.2, Dec. 1979 Published in 1980.

FSC 1966. In: Development of Fisheries in Reservoirs–G. P. Dubey (MS).

Gammon, J.R. 1970. The effect of inorganic sediment on stream biota. U.S. Environ Protection Agency *Water Pollution Cent. Res. Ser.* 18050 DWC 12/70, 1-141.

Ganapati, S.V. 1970. Energy available relationship in natural aquatic biosystems of India. *Tropical Ecology.* 11(1): 49-68.

Ganapati, S.V. and Sreenivasana, A, 1972. Energy flow in Aquatic ecosystem o£ India, *Proc IBP Syrup,* Productivity Problems of freshwater Kaaimierz Dolvy, Poland. May 1970 WARSZAW–KRAKNW, 1972.

Ghosh, A.N. and K.V. Rao, 1969. Problems of Fisheries Development in newly constructed Reservoirs in South Bihar with particular reference to Badua Reservoir. *Proceedings of the Seminar on the Ecology and Fisheries of Fresh water*

Reservoirs. November 2 7-29. Sponsored by: I CAR, Krishi Bhavan, Mew Delhi at CIFRI, Barrackpore, West Bengal.

Ghosh, Manas, 2002. No share and share alike. *The Sunday Statesman*, 19.9.2002.

Gopalakrishnan, V. and Pal, R.N. 1969. Spawning of carps in some reservoirs of India. Proceedings of the seminar on the ecology and Fisheries of Freshwater Reservoirs: November 27-29. Sponsored by: ICAR, Krishi Bhavan, New Delhi–110 001, At CIFRI, Barracknore, West Bengal.

Gopalakrishnan, V., R.N. Pal and P.K. Chakraborty, 1966. Observations on breeding of major corps in Tilaiya and Panchet reservoirs. *Cent. Inland Fish. Reso Instt* Barrackpore No. 9: 1-17.

Govind, B.V. (MS). Cage culture of common c«rp and Silver carp in Sankery Tank in Bangalore (Karnataka).

Govind, B.V., *et al.*, 1988. Culture of Catla (*Catla catla*) in floating net cages. *Mysore J. Agri. Sci*: 22: 515-522.

Govind, B.V. and M.A. Khan 1969. Prerecruitment studies in Tilaiya, Reservoir (Bihar) in context of stocking. Proceedings of the seminar on the Ecology and Fisheries of Freshwater Reservoirs. November 27-29, Sponsored by: ICAR, Krishi Bhavan, New–Delhi–1100 01, at CIFRI, Barrackpore, West Bengal.

Gulbadamov, S.B. 1961. Improvement of fishing Techniques in Inland Reservoirs of India. *Report to the Govt, of India* No, 1342 FAQ EPTA: 1-13.

Gulbadamov, S.B. 1962. Improvement of Fishing Techniques in Inland Reservoirs. *Report to the Govt. of India* No. 1944, FAO, EPTA: 1-37.

Heard, R. William 1959. The use of Hobbled gill nets in a commercial fishing of lake Carl Black well. Okhahoma. *Thirteenth Annual Conference of South-Eastern Association of Game and fish Commissioners*; 90-95.

Hey, D. 1952. *The Culture of Freshwater Fish in South Africa. Inland Fish. Dept.*, Cape of Good Hope. IIIed, p. 126.

Hora, S.L. (1945). Analysis of factors influencing the spawring of carps. Symposium on the factors influencing the spawning of Indian Corps. *Proc. nat. Inst. Sci. India.* (B). 11: 303-312.

Huet, M. 1960. General Principles of Pond construction and management, productivity and stocking. *ETAP. FAO Lecture at the III International Training Centre, Bagore. Indonesia* 1955.

IIMA 1983. In: Development of Fisheries in Reservoirs–G.P. Dubey (MS), National Commission on Agriculture.

Jhingran Arun G. 1986. Reservoir Fisheries Management in India. CICFRI, Bulletin No.45. ICAR, Barrackpore, West Bengal.

Jhingran, A.G. 1988. Reservoir Fisheries management of India. Bull 45, Central Inland Capture Fisheries Research Institute Barrackpore, pp. 68.

Jhingran, A.G. 1990. Recent advances in the Reservoir Fisheries Management in India. In: Sena De Silva (Ed) Reservoir Fisheries of Asia. Proceedings of the 2nd Asian Reservoir. Fisheries workshop held in Hongzhou Peoples Republic of Chia, 15-19 October.

Jhingran, V.G. and A.V. Natarajan, 1969. An assessment of fisheries of the DVC reservoirs in relation to stocking. *Proceeding of the Seminar on the ecology and Fisheries of Freshwater reservoirs*. November 27-29. Sponsored by: ICAR, Krishi Bhavan, New Delhi–110001 at CIFRI, Barrackpore, West Bengal.

Jhingran, V.G. and S.D. Tripathi, 1969. A review of the measures adopted for the development of the fisheries of Reservoirs in India. *Proceedings of the Seminar on the ecology and Fisheries of Freshwater Reservoirs*. Nov. 27-29.

Jhingran, V.G. 1975. *Fish and Fisheries of India*. Hindustan Publishing Corporation (India) 6–U.B. Jawahar Nagar, Delhi–110007.

Jhingran, V.G. 1982. *Fish and Fisheries of India*. Rev. ed. Hindustan Publishing Corporation (India), Delhi 666 p.

Joshi, H.C. 1990. Water pollution problems in Indian Reservoirs, *Proceedings of the National workshop on Reservoir Fisheries*. 3-4 January. Special Publication No.3. Asian Fisheries Society. India Branch.

Joshi, H.C. and P.K. Sukumaran, 1991. Water pollution investigation in the river Tungabhadra near Harihar, Karnataka. *Indian. J. Anim Sci*. 61 (2): 230-234.

Konar, S.K., T.K. Ghosh, and M. De. 1990. Management of Complex Pollutional problems of Indian Reservoirs to increase fish yield (abstract). *National Workshop on Reservoir Fisheries*, January 3-4. Asian Fisheries Society, Indian Branch, Central Inland Capture fisheries Research Institute, Barrackpore, West Bengal.

Krishnapurthy, K.N., B.V. Govind and G.K. Bhatnagar, 1964. Experimental Gill-net fishing in Tunbabhadra Reservoir. *Indian J. Fish*. 11 (1): 465-479.

Krishnamurthy, K.N. 1969. Commercial Fishing methods and experimental fishing operations in artificial impoundments. *Proc. of the seminar on the ecology and fisheries Fresh water Reservoir*. Nov.27-29, 1969 sponsored by ICAR, New Delhi at CIFRI, Barrackpore–743101, West Bengal.

Kumariah, P. and S. Parmeswaran, 1985. Design of a Circular Cage for fish culture. *J. Inland Fish Soc. India*. 17 (1 and 2): 78-80.

Kumaraiah, P., S. Parmeswaran and P.K. Sukumaran, 1986. Culture of *Tilapia Oreochramis mossambica* (Peters) in Cages. *Proc. Nat. Symp: Fish and Env*. 145-147 (Eds). B. D. Joshi. A. K. Chopra and Dinesh Bhat, Dept. of Zoology, Gurukula Kangri Viswavidyalay, Hardwar.

Kumaraiah, P., S. Parmeswaran and N.M. Chakraborty, 1991. New perceptions in cage fish culture with reference to the growth and production of Silver Corp (*Hypophthalmichthys Val*) in cages. *Proc. Nat Symp. Freshwater Aquaculture* CIFA, Bhubaneswar, 246-247.

Langlois, T.N. 1941. Two processes operating for the reduction in abundance or elimination of fish species from certain types of water areas. *Trans. North. Amer Wild Conf.* 6: 189-201.

Malhotra, J.C., K.K. Ghosh *et al.*, 1966. Report on fish spawn prospective investigations.

Mukherjee, A.B. 1990. Wave dissipating structures for protection of floating net cages in reservoirs. In: Jhingran Arun G and V. K. Unithan (eds). Reservoir Fisheries in India, Proceedings of the National Workshop on Reservoir Fisheries, 3-4 January, 1990. Special Publication 3, Asian Fisheries Society, Indian Branch, Mangalore, pp. 33-38.

Murugesan, V.K. and P. Kumaraiah, 1978. Fourth workshop report of All India Coordinated Project on Air Breathing Fish Culture, CIFRI, Barrackpore, 12-13 December.

Natarajan, A.V. and V. Pathak, 1983. Pattern of Energy flow in freshwater tropical and sub-tropical impoundments Bull No. 36. Central Inland Fisheries Research Institute, Barrackpore, pp. 37.

Nancy, R.J., C.J. Atchimson, R.V. Bulkley, B.W. Menzal, L.C. Perry and R.C. Summerfelt 1979. Effects of suspended solids and sediment on reproduction and early life of warm water fishes. A review, EPA-600/3-79-0, Corvallis, *Environmental Research Laboratory* Corvallis, 100 pp.

Natarajan, A.V. and V. Pathak, 1980. Bio-energetic approach to the productivity of man-made lakes. *J. Inland Fish Soc. India.* 12 (1) 1-13.

National Commission on Agriculture. (NCA) 1976. In: Development of Fisheries in Reservoirs–G. P. Dubey (MS).

Nuttal, P.M. and E.D. Bielby 1973. The effects of China clay on stream invertebrates. *Environ. Pollut.* 5: 77-86.

Panwar, R.S., K. Chandra, D.N. Singh, R.N. Seth and D. Kapoor. 1979. Studies on pollutional effects of Industrial wastes on Rihand Reservoir ecosystem. Proceedings of the sysaposium on environmental Biology, Muzaffarnagar 1978 (Technical papers/Eds.) S.R. Verma *et al.* Muzaffarnagar. Academy of Environmental Biology, pp. 465-479.

Prem Narayan, Kathuria, O.P. and H.V. L. Bathla, 1985. Status of Forestry Statistics in environment management. *National Seminar on Pollution Control and Environmental Management* March 16-18. Vol.V. Environmental Impact Assessment. Organised by NEERI Nagpur and Indian Association for water Pollution Control, Nagpur, Co-sponsored by: Department of Environment, New Delhi.

Raj. B.S. 1941. Dams and Fisheries; Mettur and lessons for India. *Proc. Indian Acad Sci.* 14B(5): 341-58.

Rajagopal, K.V. 1969. Biology of some commercial fishes of the Tungabhadra Reservoir with particular reference to utilization of available food, reproduction and growth. *Proceedings of the seminar on the Ecology and Fisheries of freshwater Reservoirs.* November 27-29. Sponsored by ICAR, Krishi Bhavan, New Delhi–110001 at CIFRI, Barrackpore, West Bengal.

Rajan, S. 1969. Ecological studies of the Pykara River, S. India during to pre-impoundment phase. National Institute of Oceanography, Evnakulam. *Proc. of the Seminar on Ecology and Fisheries of Fresh water Reservoir.* Bov. 27-29. 1969. Sponsored by ICAR, New Delhi at CIFRI, Barrackpore–743101, West Bengal.

Raghunathan, V. and E.R. Venkataswamy 1967. On the efficiency of Mettur Rangoon Nets in Bhavanisagar. Reservoirs. *Madras J. Fish* 3: 46-60.

Ramchandra Menon, V. 1969. On the mass occurrence of fertilized eggs of major corps in the top waters of Bhavani Sagar Reservoir. *Proceedings of the Seminar on the Ecology and Fisheries of Freshwater Reservoirs.* November 27-29. Sponsored by ICAR, Krishi Bhavan, New Delhi–110001 at CIFRI, Barrackpore, West Bengal.

Ranganathan, V. and V. Natarajan, 1969. Fisheries of Mettur Reservoir an artificial impoundment on the river Cauvery. *Proceedings of the Seminar on the ecology and fisheries of Freshwater Reservoirs* November–27-29. Sponsored by: ICAR, Krishi Bhavan, New Delhi–110001 at CIFRI, Barrackpore, West Bengal.

Ray, P. and R.K. Banerjee 1969. Hydrological studies of Tungabhadra Reservoir. *Proceedings of the seminar on the Ecology and Fisheries of Fresh-water Reservoirs.* November 27-29. Sponsored by ICAR, Krishi Bhavan, New Delhi–110001, at CIFRI, Barrackpore, West Bengal.

Reid, G. K. 1961. *Ecology of Inland waters and estuaries* Reinhold Publishing Corporation, New York: 375 p.

Samraj, P., R. Ragupathy, B. Haldorai and C. Henry 1985. Measures for the environmental protection of Nilgiri hills–An ecological perspective. National *Seminar on Pollution Control and Environmental Management,* March, 16-18. NEERI, Nagpur Vol.V. Environmental Impact Assessment. Organised by: NEERI, Nagpur and Indian Association for water pollution control, Nagpur, Co-sponsored by: Department of Environment, New Delhi.

Singit, G.S. (MS) 1983. Fish and seed production in Pens in Tugabhadra Reservoir–An Evaluation.

Singit, G.S.P. Rammohan Reddy, and H.V. Krishnamaraju 1985. Corp fry rearing in Cages. *Fishing Chimes.* 5 (5): 50-55.

Smith, P.W. 1971. Illinois streams a classification based on their fisheries and an analysis of factors responsible for disappearance of native species. *III. Wat. Hist. Surv. Biol.* Notes 76, p. 14.

Sreenivasan, A. 1969. The hydrobiological features, Primary production and Fisheries of Stanley Reservoir, Mettur dam for the years 1958-1965. *Madras J. Fish.* 5: 1-18.

Sreenivasan, A. 1972. Energy transformations through primary productivity and Fish Production in some tropical Freshwater impoundments and ponds. In: Proceedings of the IBP Unesco. Symposium on Productivity problems of Freshwaters, held in Kazimierz, Dolny, Poland, 6-12 May 1970. Edited by Z-Kajak and A. Hill bricht Itkowska, Warsawa, PWN. Police Scientific Publishers, pp. 505-514.

Sreenivasan, A. 1978. Limnology and Fisheries Management. Lectures presented at the fifth FAO/SIDA Workshop on Aquatic Pollution in relation to protection of living resources. Scientific and Administrative asis for Management measures. Manila, Philippines, 17-27 February, FAO of the United Nations Rome.

Sugunan, V.V. 1995. *Reservoir Fisheries of India*. FAO Fisheries Technical paper 345.

Suri Kavita 2006. Dal is shrinking. In: *The Statesman*, 16-3-2006.

Suri Kavita 2007. Metals ravage disappearing Dal. In: The Statesman (Kolkata) 14.2.2007.

Swaminathan, M. S. 1973. "Our Agricultural feature" Sardar Patel Memorial Lecture. Air-India International Centre, New Delhi. In: National Seminar on Pollution Control and Environmental Management. 16-18 March 1985 NEERI, Nagpur Vol.V. Environmental impact Assessment. Organised by NEERI, Nagpur, and Indian Association for water pollution control, Nagpur. Co-sponsored: Department of Environment, New Delhi.

Tamhane, R.V., T.D. Biswas, B. Das and G.C. Naskar, 159. Effect of intensity of rainfall on the soil loss and run off. *J. Ind. So. Soil Sci.*, pp. 231-238.

Tripathy, Y.R. and K.N. Krishnamurthy 1964. Experimental Fishing at the Mettur Dam, Cauvery River. *Proc. Nat. Acad. Sci. India* (B) 34(2): 87-99.

Von-Brandt, A. 1955. The efficiency of drift nets. *Journal du Conseil*. 21 (1): 8-16.

Znamensky, Yn. 1967. FAO/ETAP. Report No. 2290: 1-54.

Index